"EXTRA-ORDINARY" ERGONOMICS

How to Accommodate Small and Big Persons, the Disabled and Elderly, Expectant Mothers, and Children

"EXTRA-ORDINARY" ERGONOMICS

HOW TO ACCOMMODATE SMALL AND BIG PERSONS, THE DISABLED AND ELDERLY, EXPECTANT MOTHERS, AND CHILDREN

KARL H. E. KROEMER

**HFES Issues in
Human Factors and Ergonomics Series**

Volume 4

Supervising Series Editor

Jefferson M. Koonce

Published in cooperation with the Human Factors and Ergonomics Society, P.O. Box 1369, Santa Monica, CA 90406-1369 USA; 310/394-1811, Fax 310/394-2410, http://hfes.org.

Taylor & Francis
Taylor & Francis Group

Boca Raton   London   New York   Singapore

Human
Factors
and
Ergonomics
Society

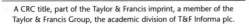

A CRC title, part of the Taylor & Francis imprint, a member of the
Taylor & Francis Group, the academic division of T&F Informa plc.

Published in 2006 by
CRC Press
Taylor & Francis Group
6000 Broken Sound Parkway NW, Suite 300
Boca Raton, FL 33487-2742

International Standard Book Number-10: 0-8493-3668-6 (Hardcover)
International Standard Book Number-13: 978-0-8493-3668-3 (Hardcover)
Library of Congress Card Number 2004062866

Library of Congress Cataloging-in-Publication Data

Kroemer, K. H. E., 1933-
    "Extra-ordinary" ergonomics: how to accommodate small and big persons, the disabled and elderly, expectant mothers, and children / Karl Kroemer.
        p. cm.
    Includes bibliographical references and index.
    ISBN 0-8493-3668-6 (alk. paper)
    1. Human engineering. 2. Engineering design. I. Title.

TA166.K773 2005
620.8'2--dc22   2004062866

Taylor & Francis Group
is the Academic Division of T&F Informa plc.

Visit the Taylor & Francis Web site at
http://www.taylorandfrancis.com

and the CRC Press Web site at
http://www.crcpress.com

Table of Contents

Preface

WHY THIS BOOK?

Customarily, the designer of workplaces and homes, machines, and jobs considers *normal adults of regular size with standard abilities*. But many persons are not "normal" by virtue of their being smaller or weaker or bigger; less able to see, hear, or move about; injured; or bedridden.

In fact, we all have phases in our lives when we are below the norm in our functions and abilities for reasons of illness or accident or because of stress and fatigue. Of course, we hope that these periods will be short, but they may extend over some time. Such phases of less-than-normal adult capabilities, however, are long for individuals with disabilities, aging persons, children and teenagers, and expectant mothers.

U.S. demographics show that at least two of every five people are "extra-ordinary" in essential aspects. In this book, I strive to recognize their problems, determine how to assess their abilities and needs, and explain how to design tools and tasks, housing, and the environment to make work and life safe and easy for them.

As a rule, designing for "nonstandard" people is not difficult, and such thoughtful design commonly makes use of a complex system or simple gadget easier for all. Hence, this book is about "extra-ordinary" ergonomics.

FOR WHOM THIS BOOK?

All of us need the reminder that we are all temporarily less abled — as children, when sick, or old.

STUDENTS, THEIR INSTRUCTORS, AND RESEARCHERS

Various methods exist to specifically measure people's sizes, strengths, weaknesses, and capabilities and, from there, to judge whether there is a need for specific ergonomic accommodations. This information is of basic value to all students and their instructors in ergonomics and related behavioral sciences, emphasizing the possibilities of, and need for, more goal-directed assessments of individuals' abilities to meet the demands of everyday activities. Development of such measures is primarily the task of research-interested ergonomists.

HUMAN FACTORS ENGINEERS, DESIGNERS, AND ARCHITECTS

Design for movement is of particular importance in human-oriented engineering; the term *mobility* assumes special meaning for children, pregnant women, older people, and persons with disabilities. Furthermore, limitations in sensory capabilities or special features in terms of body size and strength (either very small or large) generate particular requirements for devising work tasks, devices and tools, environment and habitat so that they are "people-friendly."

HEALTH CARE PROFESSIONALS, SUPERVISORS AT WORK

Helping people who suffer from temporary or permanent impediments to cope with the demands at work, at home, in traffic, or in a care facility is a compelling humanitarian task. Even working with a backache can become possible through understanding support from a nurse and the supervisor at work. These professionals, together with an ergonomist, can advise the plant engineer or the architect drawing up a house, or an office building, a hospital or other care facility about specific layout features that make life easier for all.

ABOUT THE LAYOUT OF THIS BOOK

This book is about "extra-ordinary" ergonomics. We arrived at this title after many discussions between author and editorial experts. *Merriam-Webster's Collegiate Dictionary* (11th edition) defines *extraordinary* as "going beyond what is usual, regular, or customary." However, that word does not communicate the desired distinctiveness, whereas "extra-ordinary" (with a hyphen) is distinguishing, but not proper English. We overcame that puzzle by placing "extra-ordinary" between quotation marks.

One goal in writing this book was that it should be easy to read, so persons not familiar with academic jargon would peruse it and find useful information. Yet, by its nature, some information is a bit complicated, and, for deeper understanding, readers may wish to examine publications by other authors. I wanted to simplify my text references to their works, but my own notation design did not pass editorial muster and we could not use Barzun's (2000) elegant solution. Therefore, in this book we again employ the APA notation, an accepted scholarly publication practice.

The initial two chapters establish the need for "extra-ordinary" ergonomics. The first chapter deals mostly with the information that leads to the basic but simplistic idea of the so-called normal user. The second chapter illustrates various approaches to measure the characteristics, capabilities, and limitations of those who are different from the norm. Design principles — specifically those that pertain to movement and dynamics instead of static layout — are the topics of the next two chapters.

Chapters 3 and 4 provide guidelines for designing for people who differ from the average because they are smaller, weaker, or bigger. Chapter 5 discusses design for persons with disabilities, and Chapter 6 deals with designing for the aging population. Chapter 7 takes up the topic of human factors engineering for expectant mothers. Chapter 8 addresses ergonomics for children and teenagers.

The index helps to locate text passages that treat topics of interest. The references section lists more than 300 related articles and books. The appendix contains information about body sizes in different parts of the world.

THANK YOU

Several persons have been of great help. First, I am obliged to my daughters: Henrike Sangeorzan of Main Street Psychologists, Petoskey, MI; Anne Kroemer of Family Care of Illinois in Chicago, and Katrin Elbert of Johnson & Johnson, Somerville, NJ. I have also received valuable information and enjoyed helpful criticism of text drafts from Bruce Bradtmiller of Anthrotech in Yellow Springs, Ohio; Sara Czaja of the University of Miami; Claire Gordon of the U.S. Army Natick Soldier Center; Arthur Fisk of the Georgia Institute of Technology; Klaus Hinkelmann, Maury Nussbaum, Karen Roberto, and Beverly and Robert Williges of Virginia Tech; Kathleen Klinich of the University of Michigan; Rodger Koppa of Texas A&M University; John Leslie of Wichita State University; Robert Malina of Michigan State University; Kathleen Robinette of the U.S. Air Force Research Laboratories; Alexander Roche of Wright State University; and Wendy Rogers of Georgia Tech. I particularly thank Jefferson Koonce and Rani Lueder for spending much time and effort in reviewing drafts of this book, as did other anonymous reviewers.

Kathy Buchanan, Teresa Coalson, Lynette Kessinger, Kim Ooms, and especially Sandy Dalton of the Grado Department of Industrial and Systems Engineering at Virginia Tech have helped me in various ways in the preparation of the book.

I thank Lois Smith of the Human Factors and Ergonomics Society for her help and patience over several years of writing and reviewing this book. Cindy Carelli and her team at CRC helped put this book together.

CONTACT ME

This text points out general avenues and specific steps, of which I knew in February 2005, to provide ergonomic designs for "extra-ordinary" individuals and groups of people. Please contact me at *kroemer@vt.edu* and tell me about your new data and experiences and the latest practices, so that they may appear in possible future editions of this book.

I look forward to hearing from you.

Karl Kroemer

Author

Karl "Eb" Kroemer is professor emeritus, Grado Department of Industrial and Systems Engineering, Virginia Polytechnic Institute and State University (Virginia Tech), Blacksburg, VA. He has a Vor-Diplom (B.S.) degree, a Diplom-Ingenieur (M.S.) degree, and a Doktor-Ingenieur (Dr.-Ing.) degree from the Technical University, Hannover, Germany.

Before coming to Virginia Tech in 1981, Kroemer was a research engineer in the Departments of Physiology and Biomechanics, Max-Planck Institute for Work Physiology, Dortmund, Germany. Then he served as research industrial engineer in the Human Engineering Division of the Aerospace Medical Research Laboratory, Wright-Patterson Air Force Base, Ohio. In 1973, he was appointed director and professor, Ergonomics and Occupational Medicine Divisions, Federal Institute of Occupational Safety and Accident Research, Dortmund, Germany. In 1976, he became professor of ergonomics and industrial engineering and director of the Ergonomics Research Laboratory, Department of Industrial Engineering and Operations Research, Wayne State University, Detroit, Michigan.

Kroemer is a Fellow of both the Human Factors and Ergonomics Society and the Ergonomics Society (UK). He served as UN-ILO Expert on Ergonomics in Bucharest, Romania, and Bombay, India. He was honored to be a member of the Committee on Human Factors of the National Research Council, National Academy of Sciences. He has served as member of the editorial boards of several scientific journals, including *Human Factors* and *Applied Ergonomics.*

Kroemer has served on many committees, among them DIN German Industrial Standards Revision, ISO Technical Committee 159, Department of Health and Human Services Safety and Occupational Health Study Section, and HFS/ANSI Standards Committees on Anthropometry and Biomechanics, aerial passenger tramways, and on VDT workstations.

Kroemer is a Certified Professional Ergonomist. He has lectured and consulted with industry (in North and Central America, Europe and Asia) and U.S. government agencies and provides ergonomics expertise to attorneys. He has been a researcher and consultant on such topics as human body dimensions in relation to size of workstation and equipment; physical strength and work requirements. Kroemer has authored and coauthored more than 200 publications.

Foreword

A fundamental tenet of the discipline of human factors and ergonomics (HF/E) is to "know thy user." But who is the user, usually? Is there such a thing as a typical user? Can users be sorted and categorized to enable designers to accommodate the 95th percentile? Sometimes, but not always.

Within the normal course of a lifetime, one's capabilities and limitations change — from childhood through adolescence, young adulthood, middle adulthood, and into old age. In addition, within these different phases, certain events may occur that moderate one's capabilities, perhaps on a temporary basis. For example, a broken leg from an accident, a head cold that affects hearing, or pregnancy, which certainly changes one's physical dimensions. In some cases, the disabilities are more permanent because of, for instance, birth defects, illnesses, or serious accidents.

There are also multiple situations in the workplace wherein one's capabilities may be impaired or limited in some way. Consider a firefighter in a dark, smoke-filled building, a manager trying to time-share multiple tasks, or a laparoscopic surgeon working in a very constrained space. All these situations must be considered during system design if the systems are to be used efficiently, successfully, and safely.

These users vary tremendously in terms of their functional abilities. Designers should emphasize the concepts of adaptivity and customization. For example, with respect to technology design, the focus should be on interactive, not deterministic, technology that will adapt to users' needs and not merely respond to commands. Success in the development of adaptive and interactive interfaces will depend on a solid understanding of the underlying behavior and characteristics of individuals within specific use contexts.

Human factors/ergonomics research should develop the scientific knowledge base that can be used to guide design. Thus, one of the goals of HF/E research is to discover general "laws" that govern the relationships between people and systems. However, this goal is complicated by the fact that there are individual differences in critical variables that should be considered during the design process. Why is it so important to consider individual differences? Consider what Jack Adams had to say within the realm of psychological science:

Performance on tasks without concern for individual differences would reflect the processes that define general laws of behavior. Measures from individuals in these same tasks would give human variation in the processes, and this would lead to general laws that include the prediction of individual behavior, which should be the goal of psychology. (Adams, 1989, p. 16)

Substitute "human factors" for "psychology" in that quotation, and the importance of understanding the role of individual differences as they relate to human-system collaborations is clear. The specification of the relevant factors, how such factors differ across individuals, the amount of variance in performance accounted for by such factors — these are the parameters that will lead to general principles and guidelines.

A related idea is that individual differences can serve as a crucible in theory construction (Underwood, 1975). Just as rules and laws are intended to apply generally, theories typically originate as nomothetic, meaning that they are general in nature and describe average behavior or performance. However, Underwood argued that for a theory to be really general, it must be able to accommodate individual differences — that is, the defining relationships between variables must be encompassed in the theory. The same logic applies to developing models of human-system interactions and, by implication, using those models to help solve design-related problems. The challenge is to be able to specify the range of relevant variables that should be considered in the design process. If individual differences in a variable presumed important for task performance do not correspond to individual differences in performance, then the purported relationship is ill-defined.

A second major goal of the field of HF/E is translational. Scientific evidence abounds that there are individual differences in perceptual, cognitive, motoric, and anthropometric characteristics. However, there is a gap between these data and specifications that can directly influence design. It is a sign of the maturity of the field that specifications about "extra-ordinary" individuals and people can be used to guide the design process. Translational efforts such as this book are required for the expedient deployment of our science into practice.

Dr. Kroemer's book is itself "extra-ordinary." He has succeeded in conveying the issues that must be considered when designing for the wide range of potential users. Not only does he provide the basic information in his book, he also has numerous pointers to reference sources for more information and appendices that provide anthropometric data for a variety of potential user groups. Moreover, he understands that it is not enough to classify a potential user as "old" or "pregnant" or a "teenager." These labels are simply indexes. Instead, it is important to understand how to assess the specific characteristics, needs, and capabilities of the user group of interest. Thus the inclusion of "Assessment Methods and Techniques" (Chapter 2) and "Human Factors Design Principles" (Chapter 3) provide guidance for designers in an accessible form. By focusing on the "extra-ordinary," Dr. Kroemer helps to bring about what should become ordinary for everyone: usable systems.

Wendy A. Rogers & Arthur D. Fisk
Co-Directors of the Human Factors & Aging Laboratory
School of Psychology — Georgia Institute of Technology
October 2004

REFERENCES

Adams, J. A. (1989). Historical background and appraisal of research on individual differences in learning. In R. Kanfer, P. L. Ackerman, & R. Cudeck (Eds.), *Abilities, motivation, and methodology: The Minnesota symposium on learning and individual differences* (pp. 3–22). Hillsdale, NJ: Erlbaum.

Underwood, B. J. (1975). Individual differences as a crucible in theory construction. *American Psychologist, 30*, 128–134.

1 "Extra-Ordinary" Individuals and Groups of People

INTRODUCTION

Design to match the characteristics of the user is the guiding principle of ergonomics or human factors, also called *human factors engineering*. The underlying supposition — usually undeclared — is that the user is a so-called normal adult. In fact, this assumption is fundamentally flawed. For most designs, the common user population is not a unitary, homogeneous assemblage but rather a group of individuals with diverse characteristics and varying capabilities. These points are among the thoughts that Alphonse Chapanis expressed in a letter in 1993 when we discussed my ideas for a text on "extra-ordinary" ergonomics.

Ideally, we would design everything — ranging from common items used daily to complex machinery — for anyone, including people with impairments as well as regular adults. Unfortunately, "universal design" of all items for ordinary as well as "extra-ordinary" users is often impractical and expensive or not even thought of. Therefore, even everyday household items such as bleach, toiletries, and medicines can confound users of all ages and walks of life, as H. E. Hancock, Fisk, and Rogers reported in 2001.

Customarily, tasks and jobs, machines and tools, workplaces and habitats are designed for "normal" adults of regular size who have standard abilities on such dimensions as strength, mobility, and endurance. Not surprisingly, this makes life hard for those who are substantially smaller or bigger or weaker, for children and old persons, and for all who may have limitations in their abilities to see, hear, and move around. In truth, many users vary dramatically from the "ordinary" group of adults for a variety of reasons: age, pregnancy, body size, permanent disability, or a temporary impairment; all may influence user capabilities.

Certainly, we should not design exclusively for the fictitious "regular use by normal adults." According to the 2000 U.S. census, of the nearly 282 million people in the United States, about 25% were children and teenagers, while approximately 12% were older than 64 years (http://www.census.gov/main/). This translates into about 105 million "nonnormal" persons. Adding the millions of individuals with permanent disabilities and temporary impairments, and including expectant mothers and persons with unusual body sizes, makes it evident that at least two out of five people in the United States, and across the globe, need "extra-ordinary" ergonomics.

OVERVIEW

To ensure that we understand each other, it is helpful to start by defining the basic terms *ergonomics* and *human factors* and by clarifying what I mean when I speak of special populations and "extra-ordinary" persons.

Certain traits of the human body and mind are well-defined in scientific disciplines, particularly in the fields of anthropology, physiology, and psychology. Accordingly, methods and techniques are available to take specific measurements, the results of which can be used to establish what is considered within the boundaries of "normality."

DEFINING ERGONOMICS

I use the term *ergonomics* to encompass all the deliberate efforts to design the world around us so that it fits and accommodates human beings, in order to make our daily life and the performance of tasks safe, efficient, and easy. Accordingly, ergonomics is the application of scientific principles, methods, and data drawn from a variety of disciplines to the design of engineered systems in which people play a significant role.

Ergonomists draw knowledge especially from the human-centered disciplines of anthropometry, biomechanics, physiology, psychology, and systems engineering (Byrne & Gray, 2003; K. H. E. Kroemer, H. B. Kroemer, & Kroemer-Elbert, 2001; Liu, 2003). They apply that information to simple hand tools and to multiperson sociotechnical systems. Chairs and workplaces, automobiles and buildings, working hours as well as times for leisure and recuperation must all fit diverse human users. Human variability is an indispensable fact for user-oriented design.

Ergonomics comprises human engineering, human factors, human factors engineering, human-machine (human-technology) systems, biotechnology, engineering psychology, and work physiology. These and related terms indicate differences and nuances among areas of special interests within the general field of ergonomics, but they all relate to "matching the human-made world to human capabilities and limitations" (Kroemer, 1997a). For this matching task, one needs to understand both sides of the equation: what can the human do (or not do), and what are the demands on the human. If either aspect is not quantified, fitting and matching are not possible.

Task Demands on the Person

Of the abilities and demands to be matched, the *requirements on the human* appear most easily determined and best known. In reality, though, these requirements are seldom carefully defined and quantified; for example, what kind of and how much muscle strength is actually required on a job, for work in the home, or during leisure activities? Precisely how difficult is it to prepare a meal or to operate a computer? What exactly is the effort required for changing bed linen or for walking on icy steps? What are the specific demands on an automobile driver in either light or heavy traffic? Are these demands the same for a young as well as an old person, or has the old driver altered the task by not venturing out in the dark or into unknown neighborhoods? How might (or should) the ergonomist modify task requirements so that a person with impairments can do the job?

HUMAN CAPABILITIES AND LIMITATIONS

Following long-standing tradition, we commonly sort *human capabilities* — and limitations — into two domains: physical (physiological) and mental (psychological). This approach, although tempting, is overly simplistic because it breaks the innate connection between body and mind. Yet that bisection is often used because we are accustomed to categorizing related sets of knowledge as either physiological or psychological. Newer areas of research and application attempt to recombine formerly distinct subsets of information; for example, biology and mechanics become *biomechanics*. Ergonomists strive to overcome artificial compartmentalization of knowledge about the human body into traditional disciplines. Instead, they try to combine all subsets of information to serve the overall design goal of "fitting the human."

From the middle of the last century on, many ergonomics textbooks have appeared in a variety of languages. (In 2001, Megaw provided an extensive listing of recent ergonomics literature.) In the United States alone, since the mid-1990s, more than a dozen comprehensive books (not counting specialized and applied texts) appeared on the market: for example, by Bailey (1996); Bhattacharya and McGlothlin (1996); Bridger (1995); Chapanis (1996); Chengular, Rodgers, and Bernard (2003); DiNardi (2003); Fisk, Rogers, Charness, Czaja, and Sharit (2004); Karwowski and Marras (1999); Konz and Johnson (2000); Kroemer and Grandjean (1997); K. H. E. Kroemer, H. J. Kroemer, and Kroemer-Elbert (1997); K. H. E. Kroemer and A. D. Kroemer (2001a); Kroemer et al. (2001); Phillips (2000); Salvendy (1997); Tayyari and Smith (1997); and Wickens, Gordon, and Liu (1998).

DEFINING "EXTRA-ORDINARY" INDIVIDUALS AND POPULATION GROUPS

The concept of a normal person underlies standard industrial practice in North American and European countries. In that tradition, Niebel and Freivalds (1999, pp. 318, 343) defined the "normal" operator as a person "qualified, fully trained, and able satisfactorily to perform any and all phases of the work" under "customary" conditions at a pace "representative of average."

This indicates the widespread practice of designing, purposefully or unconsciously, work procedures, systems, and everyday products for "regular adults," loosely perceived as the working population in the age range from about 20 to 40 or 50 years and living in North America, Europe, Australia, New Zealand, and Japan. Most ergonomics knowledge applies to this subsample of the Earth's population (occasionally called "Western," "industrialized," or "developed"). Only scattered information about other groups is at hand.

The people in this group of young to middle-aged adults have been of predominant interest to industry and to society apparently because they are the main generators of the gross national product, and because they are important consumers and users of products and services. The anthropometry, biomechanics, physiology, psychology, attitudes, and behavior of this segment of the population are fairly well known and generally taken as the norm.

Yet, even within North America and Europe, large population groups differ from this normative adult model: pregnant women, children, and aging and old persons, as well as those with disabilities. They are of great societal concern — and they need special attention from ergonomists — because their bodies, strengths, and weaknesses are distinct from those of the norm.

Of course, there is diversity even among "normal" adult men and women in body sizes, for instance, and in physical capabilities. Yet, in general, one can design nearly any workstation or any piece of equipment or tool so that it is usable by either women or men. In some cases, adjustment features are indispensable; in other cases, one may have to provide objects of different fixed dimensions. These adjustments and ranges, however, usually are not gender-specific but are simply necessary to fit the normal variability — for instance, in size and strength — that exists among different people. (More about variability in the section on anthropometry in this chapter and in Chapter 2).

It is a curious and rather disturbing practice to design for a small (although often not well-defined) range of variability close to the average. Some misguided design rules even cater to the fictitious "average person" who is 50th percentile in every respect, apparently a distant relative of the phantom figures that are all 5th or all 95th percentiles. (More about percentiles in Chapter 3.) In reality, there are many persons who are very unusual in one, several, or many design-relevant characteristics — for example, in body size or strength. Designated efforts make objects or processes fit such specific individuals or groups of persons. Assessment of traits such as size, strength, mobility, and appropriate design approaches are discussed in Chapters 2 through 4.

Persons who are impaired by particular disabilities need special consideration by designers. Many differ from their peers in mobility and strength, posture and size, and sensing abilities. Often, the specialized design of environment and equipment can help them avoid handicaps and live a productive, satisfactory life, as discussed in Chapter 5.

Aging people form another large segment of the population in need of special attention by ergonomists. Body size and posture, physical capabilities, and psychological traits change; most deteriorate, some quickly over a short time and others slowly over decades. To accommodate the special traits of the aging during their later working years, retirement, and waning periods poses challenging yet ethically satisfying tasks to ergonomics designers (for more, see Chapter 6).

Expectant mothers are another group of persons who need specific attention. For a limited period, their size, mobility, strength, endurance, and other characteristics change; Chapter 7 discusses this group.

Children and adolescents draw everyone's attention. The younger they are, the more different they are from the norm, the adult. Specific ergonomics information is available for small children, but there is surprisingly little systematic information about teenagers, as discussed in Chapter 8.

A DAY IN THE LIFE OF MY MOTHER

The following excerpt from an article by Fisk (1999) strikingly and lovingly conveys the daily life, with its chores and accomplishments, of a healthy and active person in her 70s and how tasks, tools, and behavior differ from those of younger adults:

Let's consider my mother. Fortunately for both her and her family, she is in rather good physical and mental health.

Mom's day begins early, about 6 a.m. First, she must take the correct dosage of several medications, and then she begins her daily routine. In the bathroom, she's careful not to slip and fall as she enters the shower. She wonders about those new products that not only clean but also add a nonslip surface to the tub. As she's getting dressed and putting on her stockings, she feels lucky that she can easily bend down. She thinks about one of her friends who can't bend over to put on her own stockings; rather than worry about it, her friend uses a cane to pull up the stockings. "An older person's life is never dull. We're always solving problems that seem routine to a younger person," thinks Mom.

After getting dressed, she fixes her breakfast using the stove and microwave, with all their respective knobs and dials. Because her memory is not as good as it used to be, she checks her daily reminder list to help plan her day. Mom does volunteer work as a county zoning board member and as a hospice volunteer, so she makes a few calls to some businesses and works her way through all of those phone menus. With each call she tries to remember what the options are and whether or not an option matches what she wants to do. Even though she has a push-button telephone, sometimes she's glad when the mechanical voice says, "If you have a rotary phone, please hold" — finally she can talk to a real person.

She remembers that some of the retirees from where she last worked are having trouble working out details about medical care between the insurance company and Medicare. Fortunately, she "knows the ropes," and the forms are not hard for her to fill out. Because she does this so much, she knows many of the administrative personnel at the doctors' offices and the hospitals, which enables her to get things worked out for the retirees.

Mom is not doing badly financially, but she doesn't have enough money to afford the minimum balances required to talk to a live teller at her bank. So she does the best she can with the on-line banking software and shifts some money from savings to checking. The bank tells her how easy it is to use an ATM, but she wonders why they don't provide training on those darn things so that she doesn't have to use the software.

Grocery shopping is on her task list for the day. The drive into town has many intersections and stop signs. Her car is relatively new, so she uses extra caution and drives slowly to avoid hitting things. A few people are upset because she takes her time at the intersections, but she thinks it's better that they're upset about her being slow than about her hitting them. On her way to the grocery store, she gets on the new four-lane road. This is a new route for her, and she needs to follow all of the signs, not get lost, and prepare for the unexpected.

At the grocery store, Mom thinks how nice it would be not to have to push such a heavy cart around. She compares prices, reading labels for the nutritional value and reading the labels on some over-the-counter medications to make sure that they do not interact with her prescription medicine.

From the store Mom goes to visit one of her hospice patients, calling on her cell phone to tell the family she is on her way. Mom is a retired nurse, but at the house, she finds an array of new medical devices. If used correctly, these devices will ease the patient's suffering. If not used correctly … well, she's glad she got the proper instruction at the last hospice training meeting. But Mom wonders how folks with less medical background can get these things to work properly.

Back at home, Mom uses a makeshift carrier for her groceries that she built; otherwise, she wonders, how else would she get them into the house? When she was younger, she cut her own firewood, but as she loses her strength, even simple lifting is not so simple any more.

Mom needs to prepare a report for the zoning board meeting this evening, so she turns on her computer and thinks about how easy it was before Windows 95 made her learn all of those new commands. She accesses her word processor and prepares and prints her report. More telephone calls, and she is off to the zoning board meeting for a series of appeals on local ordinances. After the meeting, she goes home, enters her notes into the computer, and adjusts her daily planner for tomorrow. Thinking of the full day she has planned for the next day, Mom sets her home security monitoring system and goes to bed.*

DIFFERING FROM THE NORM

By habit, we have been designing simple doorknobs or complex automated teller machines for the standard adult who possesses full physical and mental capabilities. Intuitively, it is as plain as day that an older person, a child, an expectant mother, or an individual with a disability differs from that norm. However, rather than just relying on intuition, one can objectively assess an individual's body size and muscular, metabolic, respiratory and circulatory capacities, sensory traits, motivation, and capabilities to perform under stress. Such measurements provide quantitative data that constitute a proper basis for ergonomic design.

The text that follows in this chapter lists and briefly describes traditional, well-established, and widely used categories of assessment techniques (Kroemer et al., 2001c). Their results provide information by which physicians, psychologists, human factors engineers, designers, high school coaches, and nearly everybody else habitually categorize persons and groups of people. We consider it normal when the outcomes fall into common clusters: Note, however, that the boundaries of normality are generally not defined. (I discuss newer and more specialized assessment techniques, especially for those outside the norm, in Chapter 2.)

---

* From "Human Factors and the Older Adult," by Arthur D. Fisk, *Ergonomics in Design*, January 1999, pp. 8–13. Copyright 1999 by the Human Factors and Ergonomics Society. Reprinted with permission.

ANTHROPOMETRY

General curiosity and interest in the structure of human bodies other than those of one's own group developed into standardized measures about a century and a half ago. The anthropometric technique of Martin-Saller was predominant throughout the 20th century (Garrett & Kennedy, 1971; Hertzberg, 1968; Lohman, Roche & Martorel, 1988; Roebuck, 1995). As industry and marketing began to reach around the globe, ethnic and geographic variables in body size and biomechanics became of interest to designers and engineers who needed worldwide ergonomics information (Chapanis, 1975). New computerized procedures to measure and describe the human body are beginning to augment the classic manual measurement techniques (Landau, 2000).

The NASA/Webb sourcebook published in 1978 contains an exhaustive collection of anthropometric data available in the mid-1970s. Since then, numerous publications describing national, ethnic, professional, and other groupings have appeared in the literature (see the compilations by HFES, 2002; Hsiao, Long, & Snyder, 2002; Kroemer et al., 2001; Peebles & Norris, 1998; Pheasant, 1996; Roebuck, 1995). The Appendix at the end of this book is a collection of recently published anthropometric surveys from different regions of the world.

In 1990, Juergens, Aune, and Pieper divided the total world population into 20 area groups and estimated the values of 19 main anthropometric dimensions. Table 1.1 contains excerpts from these global estimates. A comparison of these estimates with actually measured data is of considerable interest.

Anthropometry of the U.S. Population

The most reliable observations of anthropometric trends come from military surveys. For example, measurements of U.S. soldiers have been minutely recorded since the Civil War. Soldiers are a subsample of the general population, but they are selected to be fairly young, healthy, and neither extremely small nor big. Thus, their body dimensions may not exactly represent the adult civilian population, although we do not expect major differences in head, hand, and foot sizes.

For about a century, anthropometrists have observed that women and men grow taller than their parents. An analysis of 22 body dimensions on female and male U.S. Army soldiers, taken in 1988, showed that the fast "secular increase" in stature seen earlier had slowed (Greiner & Gordon, 1990). At that time, it took about 20 years (instead of 10 years taken previously) to gain approximately another centimeter. Leg length was not changing appreciably. However, body weight still increased by 2 to 3 kilograms per decade. Altogether, white, black, and Hispanic soldiers in the U.S. Army showed similar changes, whereas soldiers of Asian extraction exhibited different trends, possibly because of recent immigration. Data from Japan (Kagimoto, 1990; Roebuck, 1995) also initially indicated a rapid growth in stature, but that gain seemed to slow as well.

Surprisingly, and disappointingly, no comprehensive anthropometric survey of any national population has been done anywhere on the globe. Even in the comparatively rich United States, the last large-scale survey of civilians was in the 1940s.

TABLE 1.1
Average Anthropometric Data (in mm) Estimated for 20 Regions of the Earth

	Stature		Sitting Height		Knee Height, Sitting	
	Females	Males	Females	Males	Females	Males
North America	1650	1790	880	930	500	550
Latin America						
Native Indian population	1480	1620	800	850	445	495
European and Black population	1620	1750	860	930	480	540
Europe						
North	1690	1810	900	950	500	550
Central	1660	1770	880	940	500	550
East	1630	1750	870	910	510	550
Southeast	1620	1730	860	900	460	535
France	1630	1770	860	930	490	540
Iberia	1600	1710	850	890	480	520
Africa						
North	1610	1690	840	870	500	535
West	1530	1670	790	820	480	530
Southeast	1570	1680	820	860	495	540
Near East	1610	1710	850	890	490	520
India						
North	1540	1670	820	870	490	530
South	1500	1620	800	820	470	510
Asia						
North	1590	1690	850	900	475	515
Southeast	1530	1630	800	840	460	495
South China	1520	1660	790	840	460	505
Japan	1590	1720	860	920	395	515
Australia						
European extraction	1670	1770	880	930	525	570

Adapted from *International Data on Anthropometry* (Occupational Safety and Health Series #65), by H. W. Juergens, I. A. Aune, and U. Pieper, 1990, Geneva, Switzerland: International Labour Office.

Since then, American anthropometrists have conducted three somewhat limited investigations: one is the National Health and Nutrition Examination Survey that has been running for several decades. (Visit the Web site of the National Center for Health Statistics, Centers for Disease Control, http://www.cdc.gov/nchs/about/major/nhanes/nhanes01-02.htm, for the most current information.) The apparent problems for the ergonomist trying to use NHANES data for design purposes lie in the health-directed nature of the examination, in the selection of its clients, and especially in the small set of anthropometric data actually taken.

The second study is the 1987–1988 anthropometric survey of U.S. Army personnel (Gordon et al., 1989). During those two years, 2,208 female and 1,774 male soldiers were measured, both subsets selected to match the age categories and racial and ethnic groups in the active-duty army. (The U.S. Army is the least selective of the armed services, and so its anthropometric data appear to represent the North American adult civilian population as a whole better than do earlier U.S. Air Force and Navy surveys.) A comparison with 384 male and 124 female Midwestern workers (Marras & Kim, 1993) and with NHANES adult data (growth charts appear in Chapter 8) demonstrates only small differences in the midpoints of comparable data distributions, although their tails differ (HFES 300 Committee, 2004). A table with anthropometric data portraying U.S. Army soldiers is in the Appendix at the end of this book.

The third and most recent survey relies on modern technology, computerized surface anthropometry, which should yield interesting results (Robinette et al., 2002a, 2002b). A U.S. clothing industry group measured about 10,000 American adults during 2002–2003, but unfortunately, most of the data will not be publicly available for several years. (However, the data may be purchased; contact <SizeUSA[TC]²> at info@tc2.com.)

PHYSICAL CAPABILITIES

In our attempts to understand the workings of the human body, we regularly rely on a biomechanical model that describes segment links (long bones) connected in articulations (body joints), with the segments embellished by volumes and mass properties. Engines (muscles) that span the articulations move the segments with respect to each other and in relation to an outside reference (Chaffin, Andersson, & Martin, 1999; Kroemer & Grandjean, 1997; Kroemer et al., 1997; Marras, 1999; D. A. Winter, 1990). The body is energized by the metabolic system, which transports chemically stored energy (supplied as food and drink) to the muscles, where it is transformed into physically useful energy, especially muscular work and heat. The circulatory system, powered by the heart, distributes energy, oxygen, and metabolic by-products. (See books on human physiology for more details; for example, by Astrand & Rodahl, 1986; Schmidt & Thews, 1989).

The combustion engine used in automobiles provides a useful analogy for how the human body generates and transforms energy. The engine burns fuel, a process that requires oxygen. The combustion yields mechanical energy that moves segments as well as the entire system. Combustion by-products need to be dissipated. In the human body, the bloodstream carries fuel (mostly carbohydrates and fat derivatives, and oxygen absorbed in the lungs) to the combustion sites (muscles and other organs). The flow of blood also removes metabolic by-products (particularly carbon dioxide, water, and heat) for dissipation at the lung surfaces, with most heat and water dispersed at the skin. Thus, the process that converts chemically stored energy into mechanically useful work depends on the interactive functioning of the metabolic, circulatory, and respiratory systems.

We can express the balance between energy input (via the nutrients in food and drink) and energy output through a simple equation. In this equation, energy input is depicted as I and output as O.

If there is no change in energy storage, then

$$I = O \qquad (1)$$

In this case, the output can be further subdivided into heat H generated and work W performed:

$$O = H + W \qquad (2)$$

In everyday activities of most adults, only about 5% of the energy input is converted into work, and the other 95% ultimately becomes heat. Thus, the human body is very inefficient as a motor — but it can function well as a heater.

Ability To Do Strenuous Work

Skeletal muscles work to move body segments against internal and external resistances while doing physical work. Muscles are able to convert chemically stored energy into physical work. On demand, they can increase energy generation about 50-fold from the resting level. Such large variation requires a quickly adapting supply of nutrients and oxygen and, in the process, generates large amounts of metabolic waste products. To perform such demanding physical work, our muscles need an efficient circulatory system to provide supplies and remove wastes, and our respiratory system must be able to absorb oxygen.

In Western societies, the normal daily energy consumption is between 2,000 and 3,500 Calories per day (1 Cal = 1 kcal = 1000 cal ≈ 4.2 kJ). The associated metabolic processes require the circulatory system to provide supplies and remove by-products. The effort put forth by the circulatory system, powered by the heart, is easily observed in terms of the number of heartbeats per minute. This is why we often use a presumed monotonic relationship between heart rate (also called pulse frequency), oxygen uptake, and body strain during physical work.

Measurements of heart rate or oxygen consumption can reliably assess the dynamic work (in physical terms, force exertion together with displacement) that the body performs. In a static (isometric) effort (in physical terms, force times duration), a muscle contracts and stays that way. A maintained strong contraction hinders or completely blocks the muscle's own blood supply by compression of its capillary bed. The heart strains to overcome the resistance by increasing beating rate, stroke volume, and blood pressure. But because blood flow remains insufficient, relatively little energy is supplied to the contracted muscle and consumed there.

The ability to perform physical work varies from person to person and depends on health, fitness, training, gender, age, body size, the environment, and particularly motivation. These factors and their interactions are sketched in Figure 1.1.

It is self-evident that children, many older persons, persons with physical impairments, and pregnant women have some limitations in terms of physical ability for strenuous efforts and should not be held to standards designed to fit the normative adult model.

Muscle Strength

Skeletal muscles are similar to engines, in that they move body segments by turning limbs (links) in their intermediate joints. There are more than 200 skeletal muscles

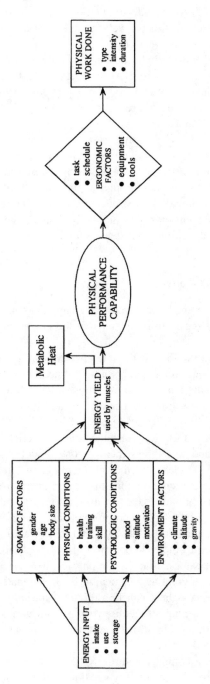

FIGURE 1.1 Traits that determine a person's capacity for physical work.

in the human body. They consist of bundles of muscle fibers that are penetrated and wrapped by connective tissues. The connective tissues embed nerves that control muscle activity and vessels for blood circulation. At each end of the muscle, the tissues combine to form tendons that attach to bones. When a muscle contracts, it applies force to the bones via the tendons.

Muscle strength is a diffuse term because it may refer to static force, which does not bring about limb movement, or to dynamic energy, which rotates limbs about a joint. The classic technique to assess body strength employed static (isometric) muscular efforts. As an example, Figure 1.2 shows static horizontal push forces of proverbial normative adult men.

Muscle strength depends not only on muscle mass and how skilled we are in using the muscle but also on our motivation (a vivid illustration of how physique and psyche interact — for more, read the book by Floru and Cnockaert, 1991). Direct effects of exercising muscles through actual exertion of strength, use, and training of these muscles include increased muscle mass, enhanced strength, and improved skill. Exercising muscles also indirectly furthers the capabilities of the circulatory, metabolic, and respiratory systems, all needed to support the generation of strength.

Thermoregulation of the Body

As mentioned earlier, the human body produces a large amount of heat while doing physical work. That heat must be dissipated into the environment, which may be quite a challenge.

Four physical factors describe our thermal environment: temperature, humidity, and movement of air, as well as the temperatures of nearby surfaces that exchange energy with us through radiation. The combination of these four components physically determines the climate, how we perceive it, and how we are able to work effectively within it.

Various techniques are in use to assess the combined effects of some or all of these four components and to express them in one model, chart, or index. The Effective Temperature (ET) index describes different climates that "feel the same." Figure 1.3 shows how we perceive combinations of temperature (measured with a dry bulb thermometer) and humidity. In slow air movement, while doing physically light work, so-called normal adults feel comfortable at 21 to 27°C (70 to 81°F) ET in the summer and in the range of 18 to 24°C (64 to 75°F) ET in the winter. (For more information, see, for example, Kroemer et al., 1997, 2001.)

Because of its meager caloric efficiency in physical work, as already mentioned, the human body generates a great deal of heat to be disseminated into the environment. Ridding itself of excess heat can be difficult, especially under challenging climactic conditions: Performing strenuous work in hot, humid, and calm air can make eliminating excess heat difficult. By contrast, in a cold environment, excessive heat loss must be prevented.

The human body has a complex control system that maintains the deep body (core) temperature close to 37°C, with only small temperature fluctuations throughout the 24-hr day due to natural (diurnal) changes in body functions.

Force-plate[1] height	Distance[2]	Force, N Mean	Force, N SD
Percent of shoulder height		With both hands	
50	80	664	177
50	100	772	216
50	120	780	165
70	80	716	162
70	100	731	233
70	120	820	138
90	80	625	147
90	100	678	195
90	120	863	141
Percent of shoulder height		With one shoulder	
60	70	761	172
60	80	854	177
60	90	792	141
70	60	580	110
70	70	698	124
70	80	729	140
80	60	521	130
80	70	620	129
80	80	636	133
Percent of shoulder height		With both hands	
70	70	623	147
70	80	688	154
70	90	586	132
80	70	545	127
80	80	543	123
80	90	533	81
90	70	433	95
90	80	448	93
90	90	485	80
	Percent of thumb-tip reach*	With both hands	
100 percent of shoulder height	50	581	143
	60	667	160
	70	981	271
	80	1285	398
	90	980	302
	100	646	254
		With the preferred hand	
	50	262	67
	60	298	71
	70	360	98
	80	520	142
	90	494	169
	100	427	173
	Percent of span**	With either hand	
100 percent of shoulder height	50	367	136
	60	346	125
	70	519	164
	80	707	190
	90	325	132

[1]Height of the center of the force plate – 20 cm high by 25 cm wide – upon which force is applied.
[2]Horizontal distance between the vertical surface of the force plate and the opposing vertical surface (wall or footrest, respectively) against which the subjects brace themselves.

*Thumb-tip reach – distance from backrest to tip of subject's thumb as arm and hand are extended forward.
**Span – the maximal distance between a person's fingertips when arms and hands are extended to each side.

FIGURE 1.2 Maximal static push forces (in N) exerted by men standing in various positions. (Adapted from "Horizontal Static Forces Exerted by Men Standing in Common Working Positions on Surfaces of Various Tractions," by K. H. E. Kroemer and D. E. Robinson, 1971, Aerospace Medical Research Laboratory Technical Report; "Man-Systems Integration Standards (Revision A)," NASA, 1989, NASA-STD 3000.)

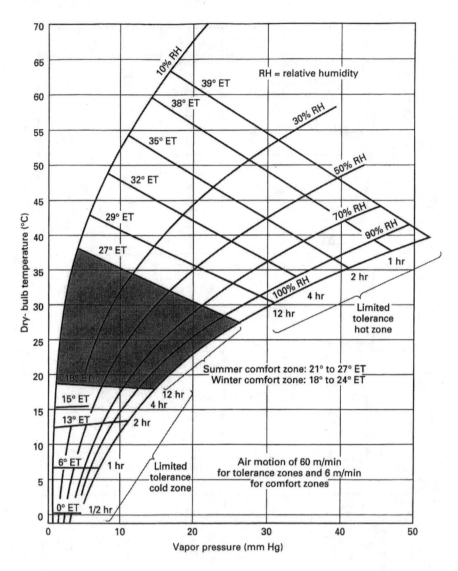

FIGURE 1.3 Indoor zones of climate comfort and tolerance for appropriately dressed sitting persons doing light work. (Adapted from "Human Factors Engineering Design for Army Material (Metric)," U.S. Army, 1981, MIL HDBK 759.)

Heat is produced in the body's metabolically active tissues, primarily the skeletal muscles, but also in internal organs and nerve tissues. Heat energy is circulated throughout the body by the bloodstream, which the heart keeps in motion. Locally, blood vessels modulate blood flow by constriction, dilation, and shunting. Heat exchange with the environment takes place at the surfaces of the skin and in the lungs.

In a cold environment, the body conserves heat automatically by reducing blood supply to the skin. This lowers the skin temperature, which reduces the temperature gradient and thus the loss of heat to the surroundings. Purposefully increasing insulation by wearing warm clothing also diminishes heat loss.

In a hot environment, the body tries to achieve two results: disseminate body heat and prevent heat gain from the environment. It does this primarily by enlarging blood flow to the skin, which augments its surface temperature, and by increasing sweat production and evaporation. Furthermore, people can take conscious actions: choose proper clothing, create air movement along the skin, and — most effective but also most costly — cool and dry ("condition") the air around them.

Selecting our clothing, shelter, and the type of activity we do affects our body's heat exchange and our degree of comfort in the environment. These controllable factors interact with the internal regulation of our blood flow to help us achieve appropriate temperature gradients between the skin and the environment. However, the internal heat control mechanism of persons with disabilities, older persons, pregnant women, and small children is generally less efficient than that of the normative adult.

CAPABILITIES OF THE MIND

It is common practice to treat the mind as if it were physically separate from the body or somehow distinct from the complex interactions of the various systems within the body. This, of course, is fallacious: Mind and body are intricately and inextricably linked. (Note that our discussion of the mind is limited here to the ergonomic aspects; it does not address such dimensions as soul, character, or spirit.)

The human has two parallel control systems: the hormonal (endocrine) and the nervous system, both of which have many similar functions. At present, few ergonomic techniques address and employ traits of the hormonal system, whereas neuromuscular functions are familiar to human factors engineers.

Most engineering psychologists consider the human mind as a receptor, processor, and generator of information. The common hierarchical model describes *sensation* as the awareness of the presence of a certain stimulus, such as warmth. Then follows *perception*, the awareness of complex properties of stimuli. It includes the interpretation of the incoming information and thus overlaps with *cognition,* the processes by which the brain accepts and transforms, reduces, elaborates, stores, recovers, compares, and finally uses the sensory inputs. The resulting output is typically a physical response, such as pressing a control button (Fisk et al., 2004).

Traditionally, input, processing, and output are treated as sequential events following each other. Accordingly, each of these stages is analyzed and researched separately. Such independent, stage-related information is then combined into a sequential model. Ecological psychologists, in contrast, assume intimate coupling of perception and action with simultaneous rather than sequential interactions. However, current knowledge still relies almost completely on the traditional sequential system. This concept is depicted schematically in Figure 1.4: A stimulus is sensed and then processed, a decision is made, and appropriate output is generated.

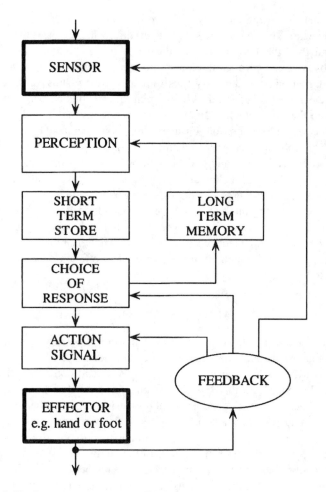

FIGURE 1.4 The human as a processor of information.

THE NERVOUS SYSTEM

The nervous system is commonly divided into three subgroups: the peripheral nervous system (PNS), the central nervous system (CNS), and the autonomic nervous system (ANS).

The afferent (sensory) part of the PNS (aPNS) collects information from internal body sensors and from sensors that react to external stimuli and transmits the information to the CNS. The CNS includes the brain and sections of the spinal cord, in which the incoming signals are processed, decisions made, and control signals generated. The efferent (motor) part of the PNS (ePNS) then transmits the control commands to body organs, especially muscles.

The ANS is responsible for general activation of the body and its emergency responses and emotions. It is generally not under conscious control. The ANS consists of the sympathetic and the parasympathetic subsystems (both containing components of the PNS and the CNS). The ANS regulates, among other involuntary functions, breathing rate, heart rate, and blood pressure and circulation, as well as the activities of smooth and cardiac muscle.

Sensors

The central nervous system receives information from internal and external receptors.

Internal receptors, called *interoceptors*, report on conditions within the body: digestion, circulation, excretion, hunger, thirst, sexual arousal, and feeling ill or sick. Internal receptors also include the proprioceptors. Among these are the muscle spindles and nerve filaments wrapped around muscle fibers, which detect the amount of stretch of the muscle. Golgi organs are associated with muscle tendons and detect their tension. Ruffini organs are kinesthetic receptors that respond to angulation of the joints. The sensors in the vestibulum report the position of the head.

External receptors, called *exteroceptors,* provide information about the interaction between the body and the outside. Most prominent among these are sensors for sound and light in the ears and eyes, discussed in more detail in the following text. Free nerve endings, corpuscles, and other receptors are located throughout the body's skin, albeit in differing densities. They transmit the sensations of touch, pressure, and pain. Because the nerve pathways from the free endings interconnect extensively, the sensations reported are not always specific for a modality; for example, very hot or cold sensations can be associated with pain, which may also be caused by hard pressure on the skin.

Injuries, especially those of the spinal cord, can deprive a person of the ability to both sense and control. Impairments in brain function, often related to injury or age, can diminish the ability to process information coming from the sensors.

Workload and Stress

Reactions to similar workloads can vary enormously from person to person and within an individual. The reaction depends on the type and magnitude of the load and on the current ability to deal with it. Any given workload may act as a stressor that generates positive stress, which spurs more activity, or it may cause "dis-stress," which overloads the person and generates ineffectiveness, evasive behavior, anxiety, and even illness (Hancock & Desmond, 2001).

The assessment of workload, whether psychological or physical, commonly relies on the so-called resource construct. This assumes that there is a given and measurable quantity of available capability, of which the job demands a portion. If the task requires less capacity than is available, a reserve exists (see Figure 1.5). Accordingly, the usual description of workload is by the portion of an individual's resource (the person's maximal performance capacity) expended in performing a given task.

A situation that demands more from a person than she or he can give generates an overload condition. This results in suboptimal task performance, and the individual is likely to suffer physically or psychologically. Conversely, an underload exists if a task demands less than full capacity. This leaves a residual capacity, the measurement

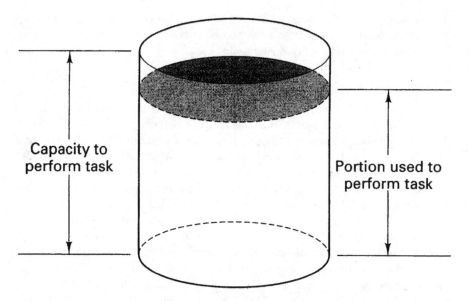

FIGURE 1.5 Traditional resource model. (From *Ergonomics — How to Design for Ease and Efficiency* (2nd ed.), by K. H. E. Kroemer, H. B. Kroemer, and K. E. Kroemer-Elbert, 2001, Upper Saddle River, NJ: Prentice Hall. Copyright 2001 by Prentice Hall. Reprinted with permission of Pearson Education, Upper Saddle River, NJ.)

of which provides an assessment of the actual workload. For more information about these measures, see the section on workload assessments in Chapter 2.

SENSING THE ENVIRONMENT

About 2,300 years ago, Aristotle established the classical five senses: seeing, hearing, touching, smelling, and tasting. In fact, touching involves several other sensations, such as force, pressure, distortion of tissues, warmth and cold, electricity, and even pain. There is also the vestibulum in the ear, which provides our equilibrium sense and our knowledge of posture. Thus, we all have at least this sixth sense.

THE VISION SENSE

The characteristics of human vision are well researched and described in the literature, which is easily accessible to ergonomists. Boff, Kaufman, and Thomas (1986) devoted seven chapters to the description of the details of human vision. The model of the "normal young adult eye" underlies the descriptions of the human visual sense; however, many of us have vision characteristics that differ from the norm, often a reduced ability. Fortunately, however, many vision deficiencies can be fairly easily remedied; for example, with artificial lenses that make many persons with afflicted vision of near-normal capability again.

To understand the phenomenon of vision, we can follow a hypothetical ray of light as it enters the eyeball, a spherical organ of about 2.5 cm diameter. Incoming light first traverses the cornea, a translucent, bulging round dome. If the eye is young and healthy, the cornea provides the refraction needed to focus on an object at least 6 meters away. Behind the cornea, the beam of light passes through a round opening called the pupil. Muscles in the surrounding tissue, the iris, adjust the size of the pupil similar to the aperture diaphragm of a camera, regulating the amount of light entering the eye. Now the light beam enters the lens. If a distant object needs to be seen, ligaments keep the lens thin and flat so that the light rays are not bent. For objects that are so close that the cornea is not sufficient to focus on them, the ciliary muscle makes the lens thicker, so that the light beams are refracted. The space behind the lens, the interior of the eyeball, is filled with vitreous humor, a gel-like fluid that has refractory properties similar to water.

Having passed through the vitreous humor, the light finally reaches the retina. This is a layer of tissue located opposite to the pupil that covers about three-quarters of the inner surface of the eyeball. It contains about 130 million light detectors. These detectors are of two kinds, named for their shapes. Most are rods, which respond even to low-intensity light and provide black-gray-white vision. There are also about 10 million cones, which respond to colored bright light. Rods and cones convert light into electrical signals that enter the brain via the optic nerve.

Mobility of the Eyes

In addition to the muscles that adjust pupil and lens, other muscles control the movement of the eyeball. Six muscles attach to the outside of the eyeball to turn it: up and down in pitch about axis y (shown in Figure 1.6); left and right in yaw about axis z; and in roll about axis x. Normally, the eyes can track a visual target continuously left and right (yaw) if it moves less than 30°/s or cycles at less than 2 Hz. Above these rates, our eyes can no longer track continuously but lag behind and then make jump-like movements (saccades) to catch up with the visual target.

Line of Sight

When our eye is fixated on a target, the line of sight (LOS, congruent with the x axis in Figure 1.6) runs from the object we are watching through the lens (pupil) to the receptive area on the retina. Thus, the LOS is clearly established within the eyeball. If we wish to describe the LOS direction external to the eye, we need a suitable reference. In the past, the horizon often served as that reference in the x–z plane (also called *medial* or *midsagittal* plane). Unfortunately, the horizon is not an appropriate reference because it does not allow for distinguishing between two distinct angles: the pitch of the eye with regard to the skull (LOSEE; see Figure 1.7) and the pitch of the head with respect to the neck, trunk, or horizon. To define the line-of-sight angle (LOSEE) suitably, we best use a reference that is attached to the skull. The ear-eye (EE) line is easy to establish and use, as shown in Figure 1.7.

Size of the Visual Target

To quantify the size of a visual target, we can define it as its length (L) perpendicular to the line of sight at a distance (D) from the eye. The target size is then expressed

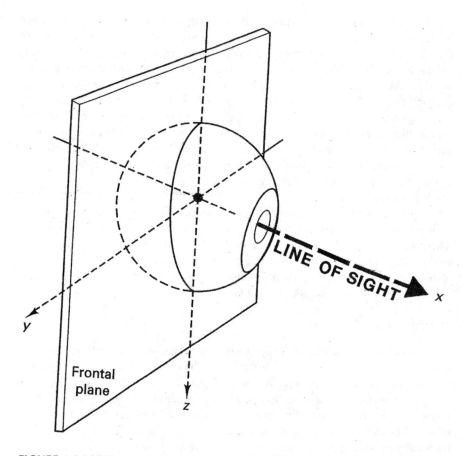

FIGURE 1.6 Mobility axes of the human eye. (Adapted from *Ergonomics — How to Design for Ease and Efficiency* (2nd ed.), by K. H. E. Kroemer, H. B. Kroemer, and K.E. Kroemer-Elbert, 2001, Upper Saddle River, NJ: Prentice Hall. Copyright 2001 by Prentice Hall. Reprinted with permission of Pearson Education, Upper Saddle River, NJ.)

as the subtended visual angle at the pupil. For small targets, the angle is $3{,}438\ L\ D^{-1}$ in minutes of arc or $57.3\ L\ D^{-1}$ in degrees. The young and healthy human eye can perceive a visual angle of at least 1 minute of arc.

Visual Acuity

The abilities to detect stationary small details and distinguish small objects are the most common ways to assess visual acuity.

A standardized high-contrast pattern at fixed distances serves to measure visual acuity. The common viewing distances are between the *far point*, 6 m (20 ft) from the eyes, and a *close point* at 0.4 m. The smallest detail that the observer can detect

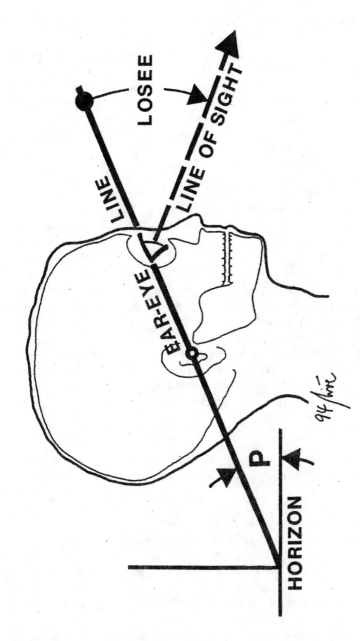

FIGURE 1.7 The ear–eye line passes through the right ear hole and the juncture of the lids of the right eye. The EE line serves as the reference for head posture: the head is held upright (erect) when the tilt angle P is about 15° above the horizon. The EE line is also the reference for the line-of-sight angle LOSEE. (From *Engineering Physiology: Bases of Human Factors/Ergonomics* (3rd ed.), by K. H. E. Kroemer, H. B. Kroemer, and K. E. Kroemer-Elbert, 1997, New York: Van Nostrand Reinhold/Wiley. Reprinted with permission of John Wiley & Sons, New York.)

or identify determines the threshold, expressed in minutes of visual arc. The reciprocal of the threshold, compared with the ability of a "perfect" eye, describes an individual's visual acuity. The most common testing pattern uses standardized black Snellen letters on a white background.

In reality, however, even people with good Snellen acuity may not be able to distinguish targets such as road signs from a busy background. As the gaze sweeps around a scene, the visual details generate a varying image on the retina, as opposed to the static laboratory test situation. Our real-world visual environment is a constantly changing array of textures of various sizes at varying distances and of differing contrasts. Not only must we focus on what is in front of us, but objects in the periphery may be of great importance as well. Consider, for example, the challenge we face driving through an intersection as a bicyclist suddenly approaches from the side.

Accommodation

Accommodation is the act of focusing on targets at various distances. Because the distance from the cornea to the retina remains constant, the actual shape of the lens itself must change so that bundles of light arriving in the eye from different distances collate in focus at the retina.

The healthy young eye can accommodate from infinity to very close distances. The farthest distance at which one can focus is called *the far point of accommodation*. The closest, the near point, is about 10 cm from the pupil for a young healthy eye, but there are large differences in focusing capabilities among individuals. Most people find it easier to focus on a close target that is in front and below the eyes rather than looking straight ahead or even upward. Optometrists have always known this and habitually place the reading section of bifocal (or trifocal) corrective lenses lowest on the lenses so that the wearer looks down at close visual targets when reading, for example, or looking at a computer screen.

Visual Problems

Aging often brings about a set of vision problems, discussed in some detail in Chapter 6. Common vision deficiencies that are usually not related to aging include the following:

- Night blindness is the condition of a person having less than normal vision in dim light — that is, with low illumination of the visual object.
- Astigmatism occurs if the cornea curves irregularly, so that an object in certain positions within the visual field does not appear sharply focused on the retina.
- Chromatic aberration exists when an eye is hyperopic for long waves (red) and myopic for short waves (violet or blue).
- Color weakness means that a person can see all colors but tends to confuse them, particularly in low illumination. Defective color vision is more common in men, of whom about 8% are color defective, compared with

less than 1% of women. Some people are colorblind, meaning that they confuse, for example, red, green, and gray. Only very few people can see no color at all or only one color.
- Floaters are small clumps of gel or cells suspended in the vitreous humor. When they lie on the line of sight, they cast a shadow on the retina.
- Cataracts are patterns of cloudiness inside the otherwise clear lens.

An ophthalmologist can alleviate some of these problems by medical or surgical interventions and by providing corrective eye lenses. Yet, many persons remain subpar in their seeing abilities. For them, human-engineering measures may be of substantial help, such as affording bright illumination and using shape-coding of objects in addition to color-coding; for example, on the lights of traffic signals.

THE HEARING SENSE

Sound is a vibration that stimulates an auditory sensation. The outer ear (auricle and pinna) collects airborne sound waves and channels them along the auditory canal to the eardrum (tympanic membrane), which then vibrates according to the frequency and intensity of the arriving sound wave. In the middle ear, the ear bones (ossicles) mechanically transmit the sound that arrives through the eardrum to the oval window.

The middle ear is filled with air; the Eustachian tube connects it to the pharynx. This tube allows the air pressure in the middle ear to become equal to the external pressure.

The inner ear is formed similar to a snail shell (cochlea) with two and a half turns. Fluid (endolymph) that fills the cochlea propagates sound waves as fluid shifts from the oval window to the round window. The motion of the fluid deflects the basilar membrane that runs along the length of the cochlea. The deflections stimulate sensory hair cells (cilia) in the organs of Corti, located on the basilar membrane. Depending on their structure and location, the Corti organs respond to specific frequencies and transmit these into electric impulses. The auditory nerve sends the signals to the brain.

A tone is a single-frequency oscillation, whereas a sound contains a mixture of frequencies. Frequencies are measured in hertz (Hz; 1 kHz = 1,000 Hz). Infants can hear high tones of up to 20 kHz, but older people can rarely hear frequencies above 12 kHz. The lowest tone humans can hear is at about 16 Hz.

The intensity of tones or sounds is measured in a logarithmic unit known as a *decibel* (dB, one tenth of a bel; see the following text). One reason for the use of a logarithmic scale is that humans perceive sound pressure amplitudes in a roughly logarithmic manner. The minimal pressure of hearing, P_o, is about 20 μPa in the frequency range of 1,000 to 5,000 Hz. The ear experiences pain when the sound pressure reaches about 200 Pa.

The sound pressure level (SPL) is the ratio between the existing sound pressure, P, and the pressure at the threshold of hearing, P_o. Thus, the definition of SPL (in dB) is:

$$SPL = 20 \log_{10}(P/P_o) \tag{3}$$

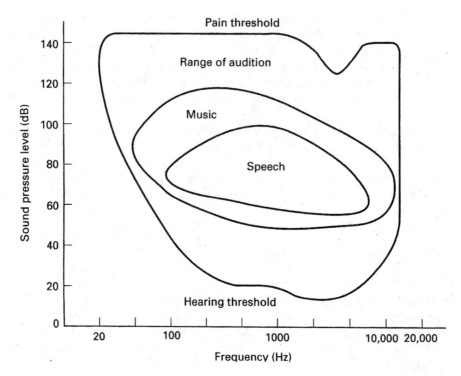

FIGURE 1.8 Ranges of adult human hearing. (From *Ergonomics — How to Design for Ease and Efficiency* (2nd ed.), by K. H. E. Kroemer, H. B. Kroemer, and K. E. Kroemer-Elbert, 2001, Upper Saddle River, NJ: Prentice Hall. Copyright 2001 by Prentice Hall. Reprinted with permission of Pearson Education, Upper Saddle River, NJ.)

where P is the root-mean-square (*rms*) sound pressure of the existing sound. With these values, the maximal dynamic range of human hearing from 20 μPa to 200 Pa is $20 \log(200/20 \times 10^6)$; that is, 140 dB.

Figure 1.8 shows the normal ranges of human hearing. These boundaries shift and tighten in persons who do not have perfect audition, such as older people or those who have suffered hearing loss. Providing signals in sounds of such frequencies and intensities that are within hearing range is generally the best human factors approach; technical hearing aids can commonly counteract individual deficiencies.

Hearing in Noise

Noise is any acoustic event that is annoying, whether loud or not. An intense sound, such as the ruckus of a construction worker's jackhammer or your favorite song blasting in your earphones, can produce a temporary threshold shift (TTS). Often, hearing eventually returns to normal, but deafeningly loud sound, even when music to your ears, may cause a permanent threshold shift (PTS), an irrecoverable loss of hearing. The severity of the threshold shift depends on the physical characteristics (intensity and frequency) of the sound and on the duration and nature of the exposure (continuous or intermittent). Damage may be immediate — for example, by a loud explosion — or may occur over time, such as with continuous exposure to intensive noise.

Impairment of the ability to hear at special frequency ranges indicates noise exposure at these frequencies. In industrialized countries, with their specific noises, noise-induced hearing loss (NIHL) usually occurs initially in the range of 3,000 to 6,000 Hz, particularly at about 4,000 Hz, and then extends into higher frequencies, culminating at around 8,000 Hz. However, reduced hearing near 8,000 Hz can also occur with aging. Hearing loss commonly brings with it difficulties in understanding speech sounds, especially of low-intensity consonants. Also, background noise, competing voices, or reverberation may interfere with the listener's ability to receive information and to communicate.

Intelligibility is the psychological process of understanding meaningful words, phrases, and sentences. The intensity level of a speech signal relative to the level of ambient noise is a fundamental determinant of the intelligibility of speech. The commonly used *speech to noise ratio* (S/N) is really not a fraction but a difference: For a speech level of 80 dB in noise of 70 dB, the S/N is simply +10 dB. With a S/N of +10 dB or higher, people with normal hearing should understand at least 80% of spoken words in a typical broadband noise. As the S/N falls, intelligibility drops to about 70% at 5 dB, 50% at 0 dB, and to 25% at −5 dB. (Other more complex assessments of speech intelligibility are described in the acoustic literature.)

Most human voices encompass a frequency bandwidth between about 200 and 8,000 Hz in speech, with men using more of the low-frequency energy than women do. Filtering or masking frequencies below 600 or above 3,000 Hz, such as occurs with many telephones, affects intelligibility only minimally. However, interfering with voice frequencies between 1,000 and 3,000 Hz drastically reduces intelligibility. A rule of thumb is that at least about 75% intelligibility is required for satisfactory communication of voice messages in noise.

People tend to raise their voices in loud surroundings and lower them when the noise subsides. In a quiet environment, men normally produce about 58 dBA, in a loud voice 76 dBA, and 89 dBA when shouting. On average, women's voice intensity is 2 or 3 dBA less at lower efforts and 5 to 7 dBA less at higher outputs. At low noise levels, people can improve communication fairly easily by raising their voices, but the ability to compensate diminishes as the noise increases. Above about 70 dBA, raising one's voice becomes inefficient, and it is really insufficient at 85 dBA or higher. Furthermore, this forced effort of shouting usually decreases intelligibility because articulation becomes distorted at extremely loud voice outputs.

Similar to medical and ergonomic measures mentioned already in relation to the vision sense, proper human engineering can alleviate detrimental effects on persons with hearing defects. Yet, for this it is essential that the existence of defects be established. Unfortunately, there seems to be a societal reluctance to admit to hearing deficiencies and to seek remedy. With the nature and severity of a hearing defect defined, countermeasures are often very effective, as described in detail in the literature (for example, by Fozard & Gordon-Salant, 2001; Haas & Edworthy, 2003).

The Senses of Smell and Taste

Smells strongly affect our ability to taste. Therefore, it is logical to treat olfaction and gustation together.

Smell

In the upper part of each human nostril, several millions of smell-reacting sensors reside in a patch of 4 to 6 cm^2, called the *olfactory epithelium*. Certain molecules trigger the sensors, which then send signals directly to the olfactory bulbs of the brain. Free endings of the fifth (trigeminal) cranial nerve are located throughout the nasal cavity. These nerve endings connect to the brain, where they generate the so-called common chemical sense: mostly irritating, tickling, and burning sensations, which initiate protective reflexes such as sneezing and interruption of inhalation (National Research Council, 1979). Most, if not all, odorants in high enough concentrations stimulate both the olfactory and trigeminal sensors.

Some substances are readily detectable even in minute concentrations, such as certain organic sulfur compounds, easily perceived as a stink even at a concentration of one molecule per one billion molecules of air. Yet, many atmospheric contaminants are odorless or very nearly so; carbon monoxide is a notorious example. Most existing odorants are mixtures of basic components, thus generating complex odors to which different people in different environments for different durations of exposure react differently. Various theories explain smell sensitivity, and two traditional systems serve to describe and distinguish the qualities of odor:

1. The *Crocker–Henderson* odor categories are fragrant, acid, burnt, and soapy (caprylic).
2. *Henning's smell prism* employs six primary odors: flowery (or fragrant), spicy, fruity, resinous, burnt, and foul (or putrid or rotten).

The effects of odors can be physiological and independent of the actual perception; or they can be perceived, psychological, or even psychogenic. Odor effects may stimulate the central nervous system, eliciting changes in body temperature, appetite, or arousal, and they may trigger potentially harmful reflexes. Unfavorable responses include nausea, vomiting, and headache; shallow breathing and coughing; problems with sleep, stomach, and appetite; irritation of the eyes, nose, and throat; destruction of the sense of well-being and the enjoyment of food, home, and the external environment; and feelings of disturbance, annoyance, or depression. Psychological and psychogenic effects concern especially one's attitude and mood, social judgments, benevolence or anger toward others, cooperation or separation, creativity, self-perception, and the ability to perform.

Taste

People can distinguish categories of taste but not easily quantifiable differences. This is partly due to the fact that the taste sense interacts strongly with the sense of smell and with texture sensations, all of which are present in the mouth. Food becomes almost tasteless if a cold impairs the sense of smell.

The traditional four fundamental tastes are salty, sweet, sour, and bitter. However, systems of other taste qualities could probably describe the human palate better. For example, more intricate taste categories might include umami, said to provide the full-bodied heft of chicken soup, red meat, aged cheese, mushrooms, and other complex foods.

We have about 10,000 taste buds (budlike collections of cells arranged in clusters), the receptors for taste. Our body replaces the taste buds continually, each bud about every 2 weeks. The buds' locations are mostly on the tongue, but they also exist in the palate, pharynx, and tonsils. The tip of the tongue seems to be particularly sensitive to sweet, the sides to sour, the back to bitter, and all to salty stimuli. Apparently, some taste buds react only to one stimulus of our four taste qualities, whereas others respond to several or all of the stimuli.

To taste a substance, it must be soluble in saliva. Certain stimuli evoke four taste qualities: Sodium chloride tastes salty, acid hydrogen ions appear sour, nitrogen alkaloids are bitter, and inorganic carbon is sweet. Temperature strongly affects perception of all stimuli. Taste sensitivity depends not only on the nature of the stimulus but also on its concentration, the previous state of adaptation to the taste, and such extraneous circumstances as smoking. Thus, taste sensitivity is highly variable and differs from person to person; some people cannot taste certain substances.

In 2004, researchers Axel and Buck received the Nobel Prize in Physiology or Medicine for their work in the late 1980s on the nature of our olfactory receptors. They defined the genes and proteins that allow us to recognize about 10,000 different odors. Their research also provided insights into the mechanisms underlying pheromone effects and taste perception. Yet, owing to the complex natures of both the human sensory capabilities and the trigger substances, diagnosed deficiencies in smell and taste abilities are still difficult to overcome. Therefore, physiological or medical treatments are often tentative. Research that clarifies the processes would be useful for assisting ergonomists in developing specific attempts to help.

THE CUTANEOUS SENSES

The sensory organs located in the skin are called *cutaneous* (*cutis* means skin in Latin) or *somesthetic* (*soma* is Greek for body). Receptors and associated sensations fall into four common groups:

- Mechanoreceptors sense touch (taction or contact), tickling, pressure, and related stimuli.
- Thermoreceptors react to warmth or cold, relative to each other and to the body's neutral temperature.
- Electroreceptors respond to electrical stimulation of the skin.
- Nocireceptors (from the Latin *nocere* for damage) sense pain.

Some researchers believe that there are no specific sensors that react to electricity or pain but that other sense organs transmit related sensations.

Sensing Touch

The taction sense reacts to touch at the skin. The term *tactile* applies if we perceive a stimulus solely through the skin, whereas the term *haptic* is appropriate when we receive information simultaneously from cutaneous and kinesthetic senses — that is, through the skin and through proprioceptors in muscles, tendons, and joints. Most of our everyday perception is actually haptic perception.

Stimuli of appropriate type and intensity trigger the tactile sensors, which generate signals that vary in frequency and amplitude and send them toward the CNS. Several types of sensors are involved. The most common type is the free nerve ending, a proliferation of a nerve that distally decreases in size and then disappears. Thousands of such tiny fibers extend through the layers of skin. They respond particularly to mechanical displacements and are very sensitive near hair follicles, where basket endings surround hair bulbs and respond to displacement of the hair shaft. In glabrous (smooth and hairless) skin, encapsulated receptors are also common; they are particularly numerous in the ridges of the fingertips.

Sensing Warmth and Coldness

We cannot discern the temperatures of objects that differ only slightly from skin temperature. This range of indistinguishable temperature differences is the *physiological zero* zone. We call a temperature below this level cool and warm when above. Slowly warming or cooling the skin beyond physiological zero may not elicit a change in sensation. This range in which no change in sensations occurs even though the temperature varies is called the *zone of neutrality*. Different body parts have different zones of neutrality.

Some nerve sensors respond specifically to cold and falling temperatures, whereas others react to heat and increasing temperatures. The two scales may overlap, which can lead to paradoxical or contradictory information. For example, spots on the skin that consistently register cold when stimulated at less than physiological zero may also register cold when they are stimulated by a warm object of about 45°C.

Temperatures that are warmer than physiological neutral are more easily sensed than those that are colder — at a ratio of about 1.6:1. We adapt to warm sensations within a short time, unless the temperatures are high. Adapting to cold takes longer and, moreover, does not seem to occur completely. Rapid cooling often causes an "overshoot" phenomenon; that is, for a short time, we feel colder than we physically are.

Sensing Pain

About 300 years ago, Descartes considered pain a purely physical phenomenon. Today, it is evident that the brain is actively involved in the experience of pain (Gawande, 1998).

Touch, pressure, electricity, heat, and cold can arouse unpleasant, burning, itching, or painful sensations. In addition to such cutaneous pain, there is visceral, tooth, head, or nerve trauma pain. The threshold for pain is a highly variable quantity, probably because pain often combines with other sensory reactions and has emotional components. One can adapt to pain, at least under certain circumstances and to certain stimuli.

Some people have experienced so-called second pain, which is a new and different pain wave that follows a primary pain after about 2 s. *Referred pain* indicates the displacement of the location of the pain, usually from its visceral origin to a more cutaneous location. An example of this is cardiac anginal pain, often felt in the left arm.

RESEARCH NEEDS

Research on sensory capabilities is still incomplete, especially for olfaction, gustation, and taction. Until the nature and functioning of human senses are clearly understood, it remains difficult — even impossible — to assess a person's sensory abilities precisely. Lacking such information, one cannot determine with certainty whether that person falls into a normal range of perception and whether special human engineering measures would improve the condition.

RESPONDING TO STIMULI

The time that passes from the appearance of a stimulus (such as light) to the beginning of an effector action (for example, movement of a foot) is the *reaction time*. The additional time required to perform an appropriate movement (for instance, stepping on a brake pedal) *is motion time* (or *movement time*). Adding the two time periods results in *response time*. (In everyday use, we often do not clearly distinguish between these terms.)

REACTION TIME

The "classic" information available on human reaction times indicates little practical time difference in reactions to electrical, tactile, and sound stimuli: They are, at best, between 130 and 180 ms. However, the reaction time following a smell stimulus is distinctly longer, taking about 300 ms. Reaction to taste takes around 500 ms, whereas it takes about 700 ms to react to the infliction of pain. These times apply under the most favorable laboratory circumstances and are for so-called simple reactions in which the individual expects a particular stimulus to occur, is prepared for it, and knows how to react to it. If there is uncertainty about the appearance of the signal, the reaction slows. If a person has to choose among several actions that can be taken, one speaks of choice reaction time, which is longer than the simple reaction time.

In real-world situations, human reactions are much slower than those measured in laboratory tests; for example, in automobile driving, simple braking usually takes a full second, whereas it takes more than 4 s in a passing maneuver (Peters & Peters, 2001).

MOTION TIME

Motion time follows reaction time. Movements may be simple, such as lifting a finger in response to a stimulus, or complex, such as moving a foot from the gas pedal to the brake pedal in an automobile. That foot movement involves not only more complicated motion elements than lifting the finger but also larger body (and shoe) masses that must be shifted, so this elaborate movement takes more time.

RESPONSE TIME

Minimizing the response time, the sum of the reaction and motion times, is a common engineering goal. The ergonomist achieves this by optimizing the stimulus and selecting the body part that is best suited to the task.

The best proximal signal is the one that is different from other signals and is received quickly, which primarily depends on modality and intensity; ideally, we might completely bypass receptors and act directly on nerves. Afferent and efferent transmission depend on the composition, diameter, and length of nerve fibers and on synaptic connections. Our best chances for reducing delays are to reduce the processing time needed in the CNS. Thus, a clear signal leading to an unambiguous choice of action is the most efficient approach to reducing delays. Choosing the most suitable body part for the fastest response includes

1. selecting a short afferent distance between sensor and CNS and a short efferent distance to quickly activate muscle, and
2. making sure that a small mass is involved; moving an eye is faster than moving a finger, and both motions are faster than moving a leg.

This discussion shows that large number of variables determine a person's reactions and associated times: the type and magnitude of the stimulus itself, its sensory perception, the processing of the signal, making a decision, and initiating and executing an action. The simple reaction time changes little with age from about 15 to 60 years but is substantially slower at younger ages and generally slows moderately as one's body ages. Of course, sensory deficiencies, reduced efficiency of the nervous system, movement disabilities, and other impairments can increase the response time beyond normal range, or may lead to incomplete or incorrect actions.

SUMMARY

The basic principle of ergonomics or human factors is to design to fit the characteristics of the user population. The common — yet often unspoken — underlying assumption is that the users are all normal adults and that the user population is not heterogeneous but is made up of individuals whose characteristics and capabilities vary relatively little from each other.

However, pregnant women, children and adolescents, older people, and persons with disabilities do differ significantly from the normative adult. Their capabilities and limitations are of specific kinds. Pregnancy makes a woman's body enlarge gradually over several months, with accompanying changes in body size and physical performance but no appreciable changes in brain and mental functions. After giving birth, her body will most likely revert to what it was before. Children, however, continuously increase in size, strength, mental ability, skill, and performance over the years until they reach adulthood. Older people follow the opposite path, unfortunately, in which essential capabilities deteriorate gradually, remaining nearly level during some periods but in others steeply declining. Persons of any age may have permanent or temporary disabilities, often of a very specific nature. Thus, all these groups have special needs that necessitate special ergonomic consideration and help.

In a systematic ergonomic approach, we can categorize and measure the traits and needs of these groups and persons. This allows us to compare the results with those that characterize the normal adult model. Familiar categories and traits that we can measure and compare include body size, mobility, strength, and performance of physical tasks. However, understanding and measuring functions of the mind and of the senses other than hearing and seeing still require research. Such information, when available, would allow the designer to accommodate the needs of special populations and "extra-ordinary" individuals.

Chapter 2 discusses specialized assessment techniques, especially those outside the norm.

2 Assessment Methods and Techniques

OVERVIEW

The fundamental goal of human-centered engineering is to match human characteristics such as body size, strengths and weaknesses, and capabilities and preferences with the relevant attributes of equipment, tasks, and procedures.

This chapter describes approaches that ergonomists take to measure individual and group characteristics, especially those that fall outside the norm. Many human traits are well defined and easily measured, such as body size, vision, or hearing. However, other design-relevant capabilities can be complex — for example, body mobility coupled with dynamic strength — and are therefore difficult to assess. Many human engineering tasks require consideration of multiple human capabilities involved in performing such complicated activities as driving an automobile or living alone in one's house.

This chapter builds on the classic assessments described in Chapter 1 but expands those methods to provide information on other than so-called normal persons.

MEASUREMENT TEAMS

In order to pair people's abilities with the demands of their various tasks, we measure human capabilities and limitations. A balanced, safe, efficient, satisfying, and harmonious match is achieved when tools, equipment, and environment fit the human.

Most traditional measurements of human traits such as body size and muscular strength are well defined, and we can express these characteristics in numbers, which the engineer uses as inputs for design — say, for locating a pedal and determining its operating force. However, many such seemingly straightforward assessment techniques are more intricate than they appear. The ergonomics literature discusses measurement methods in great detail; see, for example, the textbooks listed in the section "Human Capabilities and Limitations" in Chapter 1.

It is usually best — in fact, often necessary — to work with professionals who are experts in a particular assessment methodology, such as anthropologists, biomechanicists, gerontologists, and rehabilitation specialists. Some assessment techniques, measurement of oxygen consumption at work or evaluation of the ability to cope with stress, for example, require special knowledge and safety precautions that exceed the expertise of most human factors engineers and ergonomists. In these cases, we advise working with qualified experts including physicians, psychologists, psychiatrists, and sociologists. Proper sampling of small populations such as groupings of older people is a particularly challenging task (Kalton & Anderson, 1989).

Teamwork by professionals ensures accurate and pertinent assessments as well as the participants' well-being.

Many assessment methods concern specific traits; for instance, body size or response time. Other methods appraise the bundled abilities needed to perform interconnected functions, as in housekeeping. We discuss both types in the following text with the goal of pointing out established ways and promising new avenues.

ANTHROPOMETRIC TECHNIQUES

The major groups of methods to assess human body size rely on classic measuring rods and calipers, more recent electromechanical instruments, photography, lasers, and 3-D scanners. The newer procedures can record body contours in three dimensions, whereas the conventional technique generates a limited number of linear or curved measurements.

To measure body size in the conventional way, we use primarily an anthropometer, a graduated rod with a perpendicular sliding branch. The rod can be taken apart for transport and storage, but when put together it is usually 2 m long. (Anthropometric data are traditionally recorded in metric units.) Spreading and sliding calipers determine short distances, such as head or finger length, or the thickness of skinfolds, an indicator of body fat. A cone serves to measure the diameter around which fingers can close. Circular holes of increasing sizes drilled in a thin plate determine external finger diameter. A tape measures circumferences and curvatures. The section "Design to Fit Body Dimensions" in Chapter 4 provides more detail. (Even at the time of this writing, in 2004, almost all anthropometric information relied on data obtained in this manner. A report by Perkins and Blackwell, 1998, is the first publication to describe a study in which both classic and 3-D anthropometry assessed the body sizes of expectant mothers — see Chapter 7.)

The measurer applies these traditional instruments to the individual's body. This is simple but time consuming and invokes a major shortcoming: Most body dimensions are left spatially unrelated to each other. For example, in a side view, an individual's stature and the heights of eye, shoulder, hip, and knee are located in different yet regularly undefined frontal planes. Similarly, chest and abdominal depths occur in different horizontal planes, but their heights commonly remain unrecorded.

Attempting to describe the three-dimensional human body as a whole is a formidable task when using the collection of single data obtained via classic anthropometric techniques. This task requires cross-relating mostly distance data to recreate a 3-D body form and its contours. For this purpose, the Anthropology Branch of the Aerospace Medical Research Laboratory (the outstanding resource for anthropometric data during much of the second half of the 20th century — Green, Self, & Ellifritt, 1995) long employed a sculptor who, using an artistic concept of the human body, produced 3-D body forms based on the straight or curved distances that the anthropologists had obtained.

Photographs can depict all aspects of the human body, and provide records from which we can obtain practically infinite numbers of measurements at a time convenient for us. Pictorial records also have drawbacks, however: The equipment for data retrieval is expensive, a scale may be difficult to establish, parallax distortions occur, and we cannot palpate bony landmarks hidden under hair and skin on the record. For these and other reasons, pictorial anthropometry has not been widely employed, in spite of technical improvements such as stereophotometry with several cameras or mirrors, grids projected on the body, holography, and the use of film and inexpensive videotape instead of the initially used still photography (Froufe, Ferreira, & Rebelo, 2002; Paul & Douwes, 1993; Roebuck, 1995).

New technology allows determining the shape of the human body by recording the distances of all points on the surface from a reference; for example, with a laser. Computer programs then generate a 3-D model of the human body, from which one can extract not only the linear distances between defined surface landmarks but also any desired information about surface contours. The equipment is elaborate and expensive, but the recording process requires little exposure time from the individual. The CAESAR project, initiated in 1998, collected the 3-D body dimensions of thousands of persons in the United States, the Netherlands, and Italy (Robinette, 2000; Robinette et al., 2002a, 2002b; Robinette & Daanen, 2003).

One procedure combines features of the classic and newer techniques: Here, the measurer's hand guides a probe to the point of interest on the body, and mechanical or electronic means determine the probe's location in three dimensions (Feathers, Polzin, Paquet, Lenker, & Steinfeld, 2001; M. J. J. Wang, E. M. Y. Wang, & Lin, 2002a, 2002b). This technique is especially useful when an object partially obstructs the view, such as a wheelchair in which the person sits. The obstruction would throw off an automatic scanner, but a human measurer can locate the desired spot and place a measuring probe there.

By their nature, the data obtained by photography and 3-D scanners are not exactly the same as those measured by classical anthropometry. But the new data promise to meet the increased use of computer models of the human body in engineering design. Publications by the following researchers provide overviews of traditional and emerging anthropometric techniques: Annis and McConville (1996); Bradtmiller (2000); Burnside, Boehmer, and Robinette (2001); Feathers et al. (2001); Froufe et al. (2002); Hertzberg (1968); HFES 300 Committee (2004); Kroemer (1999b); Landau (2000); Lohman, Roche, and Martorel (1988); NASA/Webb (1978); Paquet, Robinette, and Rioux (2000); Paul and Douwes (1993); Perkins and Blackwell (1998); Robinette (1998, 2000); Robinette and Daanen (2003); Roebuck (1995); and M. J. J. Wang et al. (2002a, 2002b).

ASSESSING ENERGETIC CAPABILITIES

Especially in the mid-1900s, tables of energy expenditures were compiled for groups of occupational and athletic activities. However, actual individual efforts depend on size, health, fitness, and skill of the person as well as on specific conditions, such as the speed of action, tools used, or the prevailing climate. Furthermore, the current ways in which work is done are often different from those in prior decades when metabolic assessments were done: Forest workers use power saws now, farmers drive tractors, and housework and office tasks have changed as well. Also, the tabulated caloric costs of activities were measured on healthy adults of working age. Little is known about the related efforts on, say, teenagers or older persons.

Instead of using general tables, we can compute the total energetic cost of physical activities by adding the energetic costs of their elements, which, when combined, make up this activity. K. H. E. Kroemer, H. J. Kroemer, and Kroemer-Elbert (1997) showed how we may calculate, for example, daily energy expenditure (with 1 Cal = 1 kcal = 1000 cal ≈ 4.2 kJ)

> A person resting (sleeping) 8 hr/day, at an energetic cost of approximately 1.2 Cal/min, consumes about 580 Cal (1.2 Cal ·min^{-1} × 60 min·hr^{-1} × 8 hr). Light work for 6 hr while sitting, at 1.8 Cal/min, adds another 650 Cal to the energy expenditure. With an additional 6 hr of light work done while standing, at 2.1 Cal/min, and further, with 4 hr of walking at 2.6 Cal/min, the total expenditure during the full 24-hr day amounts to about 2600 Cal.

In many everyday activities, an essentially linear relationship exists between heart rate and energy uptake. This allows us to simply use heart rate to assess physical effort. Pulse rate is much easier to record than energy consumption, which requires measurement of oxygen uptake with complex equipment that often interferes with the activity (Webb, 1985).

Heart rate and energy expenditure provide objective information about the circulatory and metabolic requirements of dynamic work. These data, in turn, allow for statements about a task being easy or hard, as shown in Table 2.1. Such labels reflect personal judgments of what is permissible, acceptable, comfortable, or easy; the opinions of grandchildren about the "lightness" or "heaviness" of a given job probably differ from their grandparents' judgments.

Energy measurement can assess only *dynamic work* (in physical terms, force exertion together with displacement); as explained in Chapter 1, a maintained *static* (isometric) effort (force times duration, in physical terms) suppresses blood flow and increases heart rate, stroke volume, and blood pressure.

TABLE 2.1
Measures and Judgments of the Physical Demands of a Task

Energy Expended (kcal/min)	Judgments	Heart Rate (beats/min)
12.5	Extremely heavy/hard	160 or more
10	Very heavy/hard	140
7.5	Heavy/hard	120
5	Medium	100
2.5	Light/easy	90 or less

To assess a person's energetic capabilities, she or he performs standardized work, usually on a bicycle ergometer or a treadmill or while doing a step test. (Note that these standard tests are not similar to what is actually done at the workplace.) The reactions of the individual's metabolic and circulatory systems to the exercise are monitored. One liter of oxygen consumed generates approximately five Cal. In healthy adults, maximal metabolic capacities range from 5 to 10 Cal/min in continuous medium to heavy work, and the associated circulatory responses are between about 100 and 140 heartbeats (pulses) per minute.

Books written for ergonomists (including that by Kroemer, 1997b) and many physiology texts (such as Astrand & Rodahl, 1986; Schmidt & Thews, 1989) provide information about the physical load posed by work tasks and the measurement of a person's ability to perform such work. This information can guide the assessment of the physiological demands posed by today's tasks and, similarly, determine the capabilities of "extranormal" persons to meet the task requirements.

Of course, many persons with disabilities cannot perform some regular activities: Consider walking or running by a leg amputee or a frail senior. "Irregular" conditions require changes both in the test procedures and in the daily task requirements. The medical literature addresses modifications in capability assessments, and the following chapters discuss task adaptations.

ASSESSING MUSCLE STRENGTH

The term *muscle strength* allows various interpretations. For example, it may refer to force exerted with no limb movement (which is *isometric,* the most common classic test condition) or to force exerted during movements in one direction or several directions, with varying angular velocities at the joint. The force may persist over short or long periods, vary over time or be consistent, or be exerted once or repeated. In the past, descriptive terms such as *isometric, isotonic, isoinertial,* or even *static* and *dynamic* were often misused or misunderstood. Fortunately, we now have a nomenclature that defines different muscle strengths and allows us to establish protocols for their measurement (see, for example, Kroemer, 1997a, 1999a; Kroemer & Grandjean, 1997; Kroemer et al., 1997; K. H. E. Kroemer, H. B. Kroemer, & Kroemer-Elbert, 2001; Kumar, 2004; Marras, 1999).

Once we choose the strength test and the related measurement technique, we must devise an experimental protocol. This includes selecting and protecting the participants; obtaining information from them; controlling the experimental conditions; using, calibrating, and maintaining the measurement devices; and (usually) avoiding the effects of training and fatigue.

In selecting participants, we must ensure that they constitute a representative sample of the population we wish to study. Regarding the management of the experimental conditions, controlling motivational aspects is particularly difficult. In general (outside sports and medical function testing), the experimenter should not give the participant exhortations and encouragements. The following recommendations follow the so-called Caldwell regimen (Chaffin, Andersson, & Martin, 1999; Kroemer, K. H. E., Kroemer, H. J., and Kroemer-Elbert, K. E., 1997), which pertains to isometric strength testing but can be adapted for a dynamic test:

Definition: Static body strength is the capacity to produce torque or force by a maximal voluntary isometric muscular exertion. Strength has vector qualities and therefore should be described by magnitude and direction.

1. Measure static strength according to the following conditions:
   a. Static strength is assessed during steady exertion sustained for 4 s.
   b. The transient periods of about 1 s each before and after the steady exertion are disregarded.
   c. The strength datum is the mean score recorded during the first 3 s of the steady exertion.
2. Treat the participant as follows:
   a. The person should be informed about the purpose of the test and the procedures involved.
   b. Instructions should be kept factual and not include emotional appeals.
   c. The participant should be told to "increase to maximal exertion (without jerk) in about one second and then maintain this effort during a four-second count." (You may want to use a different procedure for special conditions.)
   d. During the test, the participant should be informed about his or her general performance in qualitative, noncomparative, positive terms. Do not give instantaneous feedback during the exertion.
   e. Rewards, goal setting, competition, spectators, fear, noise, and so forth can affect the participant's motivation and performance and therefore should be avoided.
3. Provide a minimal rest period of 2 min between related efforts, and more if symptoms of fatigue are apparent.
4. Describe the conditions existing during strength testing:
   a. body parts and muscles chiefly used
   b. body position (or movement)
   c. body support or reaction forces available
   d. coupling of the participant to the measuring device
   e. location and direction of the strength vector
   f. strength measuring and recording device
5. Describe the participants:
   a. population and sample selection including sample size
   b. current health; a medical examination and a questionnaire are recommended
   c. gender
   d. age
   e. anthropometry (at least height and weight)
   f. training related to the strength testing
6. Report the experimental results:
   a. number of data collected
   b. minimum and maximum values

    c. median and mode
    d. mean and standard deviation for normally distributed data points; for
       a nonnormal distribution, lower and upper percentile values such as
       1st, 5th, 10th, 25th, 75th, 90th, 95th, or 99th percentiles.

Kumar (2004) compiled the newest techniques and data pertaining to muscle strength assessments. However, as with the energetic assessments mentioned earlier, common test procedures often need to be modified to suit "extra-ordinary" people because their strengths often fall outside the ranges of their ordinary peers. The medical and especially the orthopedic literature discusses the various adaptations of strength tests.

ASSESSING MENTAL WORKLOAD

Workload may be physical or psychological, but often it is both (Floru & Cnockaert, 1991). To appraise an individual's capability to deal with the stress or strain produced by a workload, it is customary to use the *resource construct* mentioned in Chapter 1. This assumes that an individual has a measurable capability for doing the task. Depending on the nature of the job, a certain percentage of the available capability is required. If more capability is required than is available, the person is overloaded and task performance diminishes. If less capability is required than available, a reserve exists, and its measurement provides a means to assess the actual workload.

A common approach to assessing the spare capacity is to provide a concurrent secondary task. Measuring performance on the second task presumably assesses the spare capacity that remains after allocating resources to the primary task. However, if the individual allocates some of the resources truly needed for performing the primary task to the secondary task, then the second task impinges on the main task and modifies the conditions meant to be assessed.

Measurement of physiological events that occur automatically in the body, such as changes in heart rate, eye movements, pupil diameter, blink rate, or muscle tension, can often be done without affecting the primary task. However, these measures may be insensitive to the task requirements or may be difficult to interpret.

Under the hypothesis that workload changes performance, one can focus on assessing primary task execution. One nonintrusive way to do this is to measure performance on specific components of the primary task. Candidates are, for example, the participant's status of speech while doing the task, depletion of stock, disorder or clutter at the workplace, or the length of a customer queue. Such embedded measures of workload do not add to the task at hand, and hence they are neither obtrusive nor invasive.

Another approach uses subjective assessments (Floru & Cnockaert, 1991). Individual judgments of the workload rely on an internal integration of the perceived demands. They are comprehensive but may be unreliable, invalid, or inconsistent with other performance measures.

The following publications provide information about the ergonomic aspects of workload assessments that are of particular value when dealing with "extra-ordinary" individuals: Backs and Boucsein (2003); Backs, Lenneman, Wetzel, and Green (2003); Borg (2001); Caccioppo, Tassinary, and Berntson (2000); Charlton and O'Brien (2002); H. E. Hancock, Fisk, and Rogers (2001); P. A. Hancock & Desmond, 2001; Karwowski and Marras (1999); Nickel and Nachreiner (2003); Salvendy (1997); Stanton and Young (1999); Tayyari and Smith (1997); Wickens, Gordon, and Liu (1998); Wilson and Russell (2003); and the Volume 45, Number 14 issue of *Ergonomics* (2002).

ASSESSING VISION CAPABILITIES

As explained in Chapter 1, visual *acuity* is defined as the ability to detect stationary small details and to distinguish small objects. The smallest detail identified describes an individual's acuity threshold compared with the ability of a normal eye. *Visual accommodation* determines the distance D (in meters) at which we can focus on a target. The reciprocal of the target distance, $1/D$ (called the *diopter*), indicates the optical refraction needed for best focus. Thus, an object at infinity has the diopter value of 0, while a target at a 1-m distance has the diopter value of 1, as shown in Table 2.2.

The most common testing procedure uses standardized black Snellen letters on a white background. In the United States, where feet are still commonly used to measure distances, a measurement of 20/20 on the Snellen chart is regarded as perfect vision. A person is called *partially sighted* if the vision after correction is worse than 20/70 but still better than 20/200, meaning that this person can see an object at a distance of 20 ft that others can see at 70 ft or 200 ft, respectively. In the United States, a person is considered *legally blind* if the vision in the better eye after correction is 20/200 or poorer.

TABLE 2.2
Associations Between Focus Points (in Diopters) and Target Distances (in Meters)

Focus Point (diopter)	Target Distance (m)
0	Infinity
0.25	4.0
0.50	2.0
1	1
0.67	1.5
0.50	2.0
0.33	3.0
0.25	4.0
0.20	5.0

Ophthalmologists and optometrists are qualified to perform vision tests. For measurement control and standardization, they perform these examinations in laboratory settings. Yet, these settings do not reflect the actual tasks of detecting, observing, and judging visual events in often rapidly changing optical environments, such as discerning the unfolding course of a road while driving at night. Vision tests that describe a person's visual capabilities in a manner that reflects their actual use have not yet been designed. As stated in Chapter 1, developing methods to assess everyday vision capabilities realistically is a challenging but rewarding task.

The capability to recognize colors is an even more complex issue. Objectively, we describe colors primarily by their wavelengths, most commonly in the CIE trichromaticity diagram. Yet, other color-ordering classifications compete with each other for exactness and ease of use and are based on the principles of either color addition or subtraction. Still other classifications rely on appearance, perception, and judgment of colors (see Kroemer et al., 2001, for a listing and discussion). Thus, for the ergonomist, the existing complexities in theories and measurements indicate the need for developing a realistic, standardized, and useful assessment technique (Piccoli, 2001).

ASSESSING AUDITORY CAPABILITIES

Reduced ability to hear is a common occurrence in the Western world. The loss of hearing caused by environmental noise combines with age-related impairment, often making voice communication difficult for older persons. While speaking, the human voice usually encompasses a frequency bandwidth of about 200 to 8,000 Hz. Women use more of the higher frequencies than do men. Consonant sounds are more critical for understanding words than vowels. Yet, consonants generally have less speech energy than vowels and are more readily masked by ambient noise. Persons may also lose some of their abilities to speak loudly and clearly, which compounds the problem.

Direct face-to-face communication provides visual cues that enhance speech intelligibility even in the presence of background noise, whereas indirect voice transmissions (such as through the telephone) lack visual cues. Thus, when hearing-impaired persons cannot see and distinctly observe their partners, important content in the partner's speech often remains unclear. This is a severe problem, particularly for older people.

To measure hearing capability, audiometrists assess an individual's sensitivity to sounds of various frequencies and intensities in specialized laboratories with well-established testing procedures. To determine auditory thresholds, certain tones are generated at given intensities, and the listener indicates whether he or she can detect them. Such pure-tone audiometry is often combined with measures of speech comprehensibility. With the nature and severity of a hearing defect defined, countermeasures can be very effective. These may involve medication, surgery (cochlear implants, for example), hearing aids, and redesign of the environment. The literature on this is rich (see, for example, Berger, L. H. Royster, J. D. Royster, Driscoll, & Layne, 2003; DiNardi, 2003; Fozard & Gordon-Salant, 2001; Haas & Edworthy, 2003; Kroemer et al., 2001).

ASSESSING SMELLING AND TASTING CAPABILITIES

ODORANTS AND ODORS

The senses of smell and taste still pose considerable research challenges. Axel and Buck, the recipients of the 2004 Nobel Prize in Medicine or Physiology, discovered a family of some 1,000 genes that encode the many types of olfactory receptors, located on cells in a small area in the upper part of the lining of the nose. The genes appear to work in ways different from the receptor genes in our other sensory systems.

Odorants are chemical substances that can be analyzed by chemical methods. Odors are the sensations — the smells — that we derive from the presence of odorants. Most odorants are mixtures of basic components, generating complex odors to which different people in different environments who are exposed for different periods react differently. The usual test procedure is to present the substance having an odor to individuals in varying concentrations. The so-called threshold concentration is the one that induces a sensation in 50% of the trials.

Work is still under way to determine the physical and chemical determinants of odor and how we respond to them. Once that is fully understood, we may be able to predict the sensory properties of odorous materials from their chemical analysis. Perhaps we will even be able to construct an "odor meter" analogous to those used for sound and noise classification. Although this knowledge does not currently exist, various instrumental and sensory methods of measurement have been developed and applied to sources of odor and to the ambient atmosphere. Unfortunately, many of the available techniques are costly and time consuming, and there is still no consensus on any one scientific odor classification system (Hangartner, 1987; National Research Council, 1979; TRC Environmetal Consultants, 1989).

GUSTATION

The human sense of taste is only partly understood, as is the case with the sense of smell. Most of us can readily distinguish categories of taste but cannot discern quantitative differences as easily. This is partly because the sense of taste interacts strongly with sensations of smell, temperature, and texture, all of which are present in the mouth. Taste sensitivity differs from person to person. Some individuals cannot perceive certain substances (phenylthiocarbamide, for example) that most people are able to taste. Taste sensitivity depends on several interacting variables, including the nature of the stimulus, its concentration, its location in the mouth, and its time of application. Sensitivity also depends on the previous state of adaptation to the taste in question, on the chemical condition of the saliva, and on the accompanying smell.

Taste sensitivity decreases with age. The number of taste buds commonly diminishes after the mid-40s, and the remaining buds may later atrophy. Losing the ability to taste and smell can rob us of some of our former *joie de vivre*. In addition, reduced sensory ability can have grave consequences because some smells transmit important signals, such as about food singeing on the stove or, worse, a fire in the kitchen.

Publications by Ballard (1995), Hangartner (1987), and the National Research Council (1979) provide insights into the various functions of smelling and tasting.

At present, the precise capabilities of these senses are still hard to define and difficult to measure even though, as Ackerman (1990) and Schlosser (2002) vividly described, huge industries cater to them. Better knowledge would enable us to make improvements for the sensory-deprived in at least two respects: the ergonomics issue of attention-getting signals (for example, of a fire) and the matter of enhancing well-being and pleasure through smell, drink, and food.

ASSESSING TACTION SENSES

Many uncertainties hamper the study and ergonomic use of the cutaneous senses. Much of our current understanding relies on research done about a century ago, when — according to modern judgment — stimuli were rarely well defined and research procedures were less disciplined.

There is no question that certain skin sensors react to the stimuli of force or pressure and of warmth and cold, and that taction sensors are located in different densities all over the body. Yet, we still do not understand the exact functioning of sensors, tactile or haptic. Many sensors react to two or more distinct stimuli simultaneously and produce similar outputs, and it is seldom clear whether (or which) specific sensors respond to a given stimulus. The pathways of signal conduction to the central nervous system are complex and may be joined by other afferent paths from different regions of the body. The signals arriving in the central nervous system are interpreted in unknown ways there. Sherrick and Cholewiak (1986) provided an overview of the unsettled state of theories and knowledge in this field.

Classical experiments concerned an individual's ability to perceive the presence of a stimulus on the skin, to locate one stimulus or two simultaneous stimuli, and to distinguish between one stimulus and two stimuli applied at the same time. Whereas earlier research procedures were highly individual, modern research uses mostly three classes of stimulation:

- step functions, in which a displacement is produced quickly and held for a period of a second or so;
- impulse functions, in which a transient of some given waveform is produced for a few milliseconds;
- periodic functions, which displace the skin at constant or variable frequencies for several milliseconds.

A variety of transducers can impart these forms of mechanical energy to the skin. Most research has used skin displacement, but other experiments rely on units of force or on measures of energy transmitted.

Although humans can obviously sense temperature, there is no agreement on what the sensors are in our bodies. The conventional experimental procedure of applying pointed metallic cylinders of different temperatures (thermodes) to the skin is imperfect because it generates both touch and temperature sensations. Cooling or warming by air convection or radiation may be a better way to stimulate the receptors involved. An additional complication arises from the fact that temperature sensations are relative and adaptive.

Strong touch, electricity, heat, and cold can generate unpleasant, itching, and burning sensations and even pain. But at present, it is not clear whether or not modality-specific pain sensors exist. It is also questionable whether there are distinct "pain centers" in the central nervous system. Pain can range from barely felt to unbearable. The threshold for pain is a highly variable quantity, and, under some circumstances, one can adapt to pain that is not too intense. According to gate control theory, sensory signals must go through a gating mechanism in the spinal cord that either lets them pass or stops them. Individual emotions seem to control the gate, which would explain the apparent different pain thresholds and tolerances in individuals and groups of people (Gawande, 1998).

Boff and Lincoln (1988) and Sherrick and Cholewiak (1986) compiled the classic information on touch sensitivity of the human body. Our understanding of our tactile senses relies more on experience than on precise scientific information; some of the older work should be redone with today's experimental procedures. Much new research is needed to provide a solid basis for the assessment of taction sensations, which could then serve to distinguish differences in individual capabilities.

ASSESSING RESPONSE TIMES

The time from the appearance of a stimulus (such as light) to the beginning of an effector action (for example, movement of a foot) is the *reaction time*. Usually *motion time* follows, and combined, they result in *response time*.

REACTION TIME

Seemingly innumerable experiments on reaction time were conducted from the 1930s to the 1960s (Swink, 1966; Wargo, 1967). Since then, many different tables of reaction times have appeared in engineering handbooks. Some of these tables apparently were consolidated from various sources; today, the origin of many data, the experimental conditions under which they were measured, the accuracy of the measurements, and the individuals who participated are no longer known. The following list of approximate minimal reaction times (Swink, 1966) is typical of generally used but somewhat dubious information, often applied without much consideration or confidence:

- electric shock: 130 ms
- touch and sound: 140 ms
- sight and temperature: 180 ms
- smell: 300 ms
- taste: 500 ms
- pain: 700 ms

This list shows little practical time difference in reactions to electrical, tactile, and sound stimuli. The slightly longer reaction times for sight and temperature stimuli may lie well within the range of measuring accuracy or the variabilities among persons. However, the time following a smell stimulus appears distinctly longer and that for taste yet longer, whereas it takes by far the longest to react to the infliction of pain.

Time passes between the appearance of a signal on the input side at a given section of the nervous system and its reappearance on the output side. Such time delays occur at the sensor, in afferent signal transmission, in central processing, in efferent signal transmission, and in muscle activation.

Estimated time delays are, according to Wargo (1967),

- at the receptor: 1 to 38 ms
- along the afferent path: 2 to 100 ms
- in CNS processing: 70 to 100 ms
- along the efferent path: 10 to 20 ms
- muscle latency and contraction: 30 to 70 ms

Simply adding the shortest times leads to the theoretically shortest possible reaction times. In reality, there is little reason to expect a real-life situation in which all the times are shortest.

If a person knows that a particular stimulus will occur, is prepared for it, and knows how to react to it, the resulting reaction time is called *simple reaction time*. Its duration depends on the modality of the stimulus, as listed earlier, and its intensity. If conditions deteriorate, as when there is uncertainty about the appearance of the signal, the reaction takes longer.

The simple reaction time changes little with age from about 15 to 60 years, but reactions are substantially slower at younger ages and slow moderately as one grows old.

If a person has to choose among several actions that can be taken, *choice reaction time* is involved. It is longer than the simple reaction time and expands further if it is difficult to distinguish between several stimuli that are similar but if only one of them should trigger the response. The length of a choice reaction time is a logarithmic function of the number of alternative stimuli and responses. This can be expressed mathematically as

$$RT = a + b \log_2 N$$

where a and b are empirical constants and N is the number of choices. N may be replaced by the probability of any particular alternative. In that case, $p = 1/N$, and the preceding equation changes to

$$RT = a + b \log_2 (1/p)$$

MOTION TIME

Motion time follows reaction time. Movements may be simple, such as lifting a finger in response to a stimulus, or complex, such as swinging a tennis racket. Swinging the racket involves not only more intricate movement elements than lifting a finger but also larger body and object masses that must be moved (which takes more time).

Motion Time depends on the distance of the movement and its required precision. This relationship, called Fitts' law, can be stated as

$$MT = a + b \log_2(2D/W)$$

where D is the distance covered by the movement and W the width of the target. The constants a and b depend on the particulars of the situation (such as the body parts involved, the masses moved, and the tools or equipment used), the number of repetitive movements, and skill (training and experience).

RESPONSE TIME

Minimizing response time — the sum of the reaction and motion lags — is a goal of human factors. Optimizing the stimulus and selecting the body part that is best suited to the task are the obvious choices for the ergonomist. To do so usually requires both the selection of certain tasks (including equipment and procedures) and the assessment of related capabilities of the prospective user. This is often an iterative process until the best match becomes apparent.

In general, the techniques of assessing reactions and responses are well established and yield reliable information about participants' individual capabilities in the laboratory. Yet, as with other faculties already discussed, the application of those test results to everyday activities, especially of "extra-ordinary" persons whose capabilities are impaired, can be difficult. There are two main causes. One is the already repeatedly bemoaned lack of relevance of laboratory test results to real-world performance, and the other relates to the first problem: how to design a task or piece of equipment to accommodate a person who has shown deficiencies in one test — and often in several tests. This requires, in many cases, an ergonomics team consisting of a physician, psychologist, and engineer.

ASSESSING COMPLEX CAPABILITIES

Assessing the task-relevant abilities or disabilities of a person provides the knowledge necessary for devising equipment and procedures that truly fit the individual and for the selection and design of ergonomic aids. Some categories of impairments can be grouped together, such as certain vision or hearing deficiencies, and common human engineering aids can be prescribed. However, usually we must assess many impaired persons individually, instead of in groups, to establish their specific needs.

Physicians and researchers in biomechanics, orthopedics, and especially rehabilitation have refined the available measurement instruments. Other professions also employ procedures that help us understand what people do, would be able to do, or should not be required to do at work and in private life. For example, in industrial and safety engineering, job and task analyses as well as method and time studies serve as routine tools. They determine the components of activities and the nature of their elements and their sequences. With that knowledge, engineers can improve activities so that they may be done more easily, quickly, and safely (Hammer & Price, 2001; Niebel & Freivalds, 1999). The measurement of reaction and motion times mentioned earlier is another example, taken from the human engineering field, of assessing human characteristics and using this knowledge to make life easier. Combining methods and techniques from various professions and adapting them suitably can provide new approaches and novel tools (discussed later) to assess the capabilities of "extra-ordinary" persons, such as children, pregnant women, or individuals with disabilities.

ACTIVITIES OF DAILY LIVING

In the following section I discuss the development of new measurements. The aging population provides a convenient example and a topic that is especially compelling because most of us will get to that stage — if we are lucky.

Our daily lives largely consist of clusters of activities that have become routines, such as putting on clothes, brushing teeth, preparing and eating meals, reading, and writing. From childhood on, we learn these habits and skills and do not think about them until illness, injury, or aging makes them difficult or impossible to do. Many aging persons remain relatively healthy and functionally capable. In North America, for example, even at ages 85 and older, more than half the people living at home can perform daily living activities without substantial problems. But about 30% of persons older than 65 years in the United States have limited capacities for self-care and home-management activities (D. B. D. Smith, 1990). This underscores the heterogeneity of the older population. We cannot presume that at given age — say, 75 years — all, most, or many people will suffer serious deficiencies in performing certain tasks.

Given the great variation in individual capabilities of people of all ages and the variety of tasks that they perform, assessing all or even most related skills and deficiencies would pose a huge challenge. Instead, for practical reasons, researchers select certain sets of basic activities of daily living (ADL) and measure behavioral competence in these groups of activities. For instance, a sample set of five ADL (transfers, mobility, dressing, bathing, and toileting; discussed in more detail in Chapter 6) customarily serves to describe older persons' levels of activity competence.

In 1982 and again in 1990, Lawton surveyed a local convenience sample of independently living older persons and recipients of in-home services. In both observation periods, the groups spent about 3.5 hr/day watching television. Nearly one third of the waking day was spent on ADL, but most persons — especially impaired individuals — had long periods of inactivity.

In 1990, Clark, Czaja, and Weber surveyed 244 independently living persons between the ages of 55 and 93 years for ADL performance, help received, and problems encountered. They videotaped a subset of 60 persons in their homes, at a laundromat, and at a grocery store; 47 of these individuals were then visited in their homes, where the environment was observed and measured. The researchers analyzed 25 ADL tasks to identify specific demands. For example, the subtask *changing bed linens* was difficult to perform for many older persons. A component analysis indicated that the bed was too low for this task, requiring them to bend and reach more than they could. In general, shopping, cleaning, and dressing were also difficult, apparently because they involved stressful activities such as bending, leaning, reaching, lifting and lowering, pushing and pulling, and fine manipulation.

Such findings have a direct bearing on several important ergonomics issues: The first conclusion is that ADL (and even their finer subdivisions, instrumental activities of daily living [IADL] — see Chapter 6) are useful as quick and broad categories of ability assessments. However, the ADLs are not specific enough to pinpoint the embedded activity elements that actually are difficult or easy to do. The second conclusion, following from the first one, is that usually one cannot derive ergonomic recommendations for the proper design of tasks and equipment from an ADL (or an IADL) as a whole, but that one must consider its elemental component tasks.

According to the National Health Interview Survey 1984, a sample of about 13,000 males aged 55 years and older indicated (across all age groups) *fewer* difficulties with preparing their own meals than approximately 17,000 females in the same age groups (Analytical and Epidemiological Studies, Series 3, No. 25, DHHS [PHS] 87-1409, 1986, p. 35). That men have *fewer* difficulties contradicts the common experience that men, as a group, are less able to prepare meals than are women. Likely explanations for the paradoxical statement are that the men actually prepared fewer or simpler meals, or that they were more easily satisfied with their own performance than were the women. Whatever the explanation, the apparent fallacy of the findings indicates that uncontrolled self-assessments and self-reporting can lead to misleading judgments and conclusions.

SYSTEMATIC GATHERING OF INFORMATION

Information from existing aging research often does not meet the architect's or engineer's needs for specific ergonomic data. This is not surprising because the traditional goal of gerontological research was to understand the aging process (Birren, 1996; D. B. D. Smith, 1990), not to gather data for human engineering. Getting to know and understand processes is the goal of research in most areas of human physiology, psychology, and behavior, but such scientific endeavors do not directly provide design guidance.

INTRINSIC PERFORMANCE ELEMENTS

This leads to a conceptual framework for gathering quantitative information about human capabilities or limitations, which includes the following axioms:

- Sets of intrinsic performance elements (IPEs) exist that are the fundamental components of more complex activities (tasks). For example, specific elements of muscle strength, coordination, and speed, as well as sensory capabilities, memory, and the like, are the elements needed to perform activities such as dressing, eating, walking, or working with a computer.
- These IPEs require proper definition so that quantitative measurements are possible.
- Performance on IPEs determines performance of the total task. Specifically, inability to do a critical IPE leads to inability to execute the activity as planned, or to perform it at all.

In sum, IPEs are the *measurands* (Kondraske, 1990) by which one can reliably assess an individual's ability or inability to perform a task. If properly selected, performance measurements of critical IPEs allow prediction of a person's or a group of persons' ability to perform complex tasks.

The "Allied IPE" Model

Thus, we can define a task, such as an ADL, by a (preferably small) set of essential and critical IPEs. A person's performance of such a set of allied IPEs predicts success or failure of the overall activity. (Some, or many, of these IPEs will also be essential and critical components of other tasks.) Therefore, identifying sets of critical measurands, of single and allied IPEs, is the first and essential step toward a systematic assessment of capabilities (or a lack thereof) to perform a given task — and ultimately, as needed, of changing procedures and tools by suitable ergonomic design.

Sets of allied IPEs fall into different categories:

- The first division of allied IPEs is physical (anthropometric and physiological), concerning such attributes as size, reach, strength, and circulatory and metabolic functions.
- A second division of allied IPEs comprises sensory capabilities such as seeing, hearing, feeling, tasting, and smelling, all of which are associated predominantly with the afferent part of the peripheral nervous system.
- Another division of allied IPEs concerns the motor (efferent) part of the nervous system and mainly deals with the execution of physical actions such as body movements.
- A major division of allied IPEs concerns the central part of the nervous system (the mind) and its functional cognitive capabilities, such as perception of sensory inputs, information processing, decision making, judgments, and short- and long-term memory capabilities.
- Additional categories of allied IPEs may be added as needed, or different sets of allied IPEs established, to identify ergonomic design data appropriately.

Traditionally, researchers have tried to assess these traits in a variety of scientific disciplines; for example, in medicine, geriatrics, psychology, rehabilitation engineering, and sports. However, there has been no attempt so far to unite the various disciplinary procedures to form a smaller, purposefully defined, yet common set of elemental resources available to an individual or a group of people. When done systematically, identifying and assessing allied IPEs promises to be a limited undertaking that lends itself to quick and inexpensive execution.

Review of Existing Assessments

In most disciplines of physiological and behavioral research, well-established function tests exist that can be assembled into IPEs, generally with little or no modification. For a general survey, the existing techniques may be divided into groups related to scientific subsections, as follows.

Gerontology

In gerontological research, different sets of tests have been applied (Birren, 1996; Birren and Schaie, 2001; Kane & Kane, 1981; Lawton, 1990; Lawton & Herzog, 1989). The best-known and most often used assessments are the following:

- Physical health (really, lack of health) is assessed in terms of bed days, restricted activity days, hospitalizations, pain or discomfort, and a large number of associated medical and psychological assessments.
- ADL; for example, eating, bathing, toileting, dressing, and ambulating.
- IADL; for example, cooking, cleaning, telephoning, writing, and traveling.

ADL and IADL have received much attention and are frequently applied in assessments (see, for example, Clark, Czaja, & Weber, 1990; Lawton, 1990; Lawton & Herzog, 1989; D. B. D. Smith, 1990). The advantage of using ADL and IADL lies in their practicality and their ability to provide an easily understood description of physical functioning. However, they also have a number of serious drawbacks. One is their basic "lack of conceptual clarity," as Kane. and Kane (1981, p. 25) stated during the early development stage decades ago. The conceptual problems lie both in the exact nature of the physical task requirements and in the interaction of physical competence with factors such as motivation, opportunity, and environmental influences. The other fundamental problem involves self-reporting, as in the example of the ability to prepare one's own meals, mentioned earlier. Reporting by others does not easily solve the problem of variability in ratings. In the past, raters with different professional backgrounds and experience used direct observation, memory, or even secondary sources.

In spite of these criticisms, the exemplary concept of ADL and IADL is practical and can be improved and defined further by suitable allied IPEs, which can be measured in analytic, quantitative, and controlled ways.

Medicine

The physician has the important role of diagnosing health, which often includes the categorization of impairment, temporary or permanent, and especially of treating health deficiencies. For this, a plethora of medical assessments and resulting indices is available. These address more than a dozen organ systems, including cardiac, vascular, respiratory, hepatic, renal, musculoskeletal, neural, metabolic, and psychiatric aspects, as compiled in the Cumulative Illness Rating Scale, Cornell Medical Index, health index, Pace II, patient classification for LTC (long-term care), and others. Furthermore, medical practitioners use a large set of procedures to assess the specific physical capabilities of persons with temporary or permanent impairments. For instance, to assess hand impairment, tests such as the following are customary: Hand Dynamometer Strength, Tapping, O'Connor Finger Dexterity, Minnesota Rate of Manipulation, Placing, Displacing, One- and Two-Hand Turning and Placing, Pincing, and Clenching (American Medical Association, since 1958).

Many direct and indirect relationships exist among these medical measurements, and with physiological, gerontological, psychological, and biomechanical tests. Dividing them into sets of allied IPEs is facilitated by the tradition of cooperation between physicians and ergonomists.

Physiology

In physiology, particularly in occupational (work) physiology, several tests are routinely applied to assess the ability of an individual to perform certain work tasks. This assessment may either be done under standardized conditions (such as while walking on a treadmill, pedaling an ergometer bicycle, or doing a step test) or while a person performs occupational or leisure activities. The two most common tests rely on the measurement of oxygen uptake and heart rate.

As discussed in Chapter 1, the ability of the body to perform physical work fundamentally depends on the ability of muscle cells (mitochondria) to transform chemically bound energy into mechanical energy for muscular work. These energy-yielding processes depend, in large part, on the capacities of "service" functions; on the circulatory system to deliver fuel and oxygen to the working muscle fibers and to remove metabolic by-products; on the oxygen uptake capacity in the lungs; on the cardiac output; and on the nervous and hormonal systems that regulate these functions. Many of these functions depend, in turn, on somatic factors, health and fitness, age and gender, and muscle mass and body dimensions. Yet, physical performance is also dependent on environmental conditions and the will to mobilize one's resources for the accomplishment of the task in question (Astrand & Rodahl, 1986).

The traditional partnership between human factors engineers and physiologists is similar to that between ergonomists and physicians. In fact, many medical measurements are basically physiological and mechanical tests. Thus, their assembly into sets of allied IPEs should be a fairly straightforward task.

Psychology

There are myriad psychological tests described in texts, including the continuing series of mental measurements yearbooks (MMY, first published in 1938 by Buros), the *Test Validity Yearbook* by Landy (first in 1989), and the American Psychological Association (since 1953). One may classify the tests by administrative or behavioral categories; a division by content yields the following results:

- intelligence tests, mechanical aptitude tests, ability tests (sensory or motor)
- personality and interest inventories
- multiple aptitude test batteries
- work samples
- interviews (unstructured and structured).

Of special interest in this context are sensorimotor or psychomotor tests developed during the last few decades. They include control precision, multilimb coordination, response orientation, reaction time, speed of movement, manual and finger dexterity, aiming, and many others (Fleishman & Quaintance, 1984; R. V. Smith & Leslie, 1990). Many of these tests are, or can become, allied IPEs because psychology and human factors are intensively interwoven.

Biomechanics

Biomechanics is a fairly new discipline that applies mechanical principles and procedures to the human biological system. Biomechanics relies heavily on physiological and anthropometric information.

Human muscle strength — a typical topic in physiology, now in the domain of biomechanics — can be assessed, in terms of force or work applied by the body to an outside object, using load cells, strain gauge equipment, and other instruments that are either commercially available or specially designed. Measuring the efforts of individual muscles and their interplay is a more involved task. Muscle interactions create complex internal strains, particularly torques around body joints and compression forces in the joints.

Current biomechanical modeling emphasizes the spinal column, specifically the lumbar section because of the all-too-common low back pain (Chaffin et al., 1999; Marras, Davis, Ferguson, Lucas, & Purnendu, 2001; Nordin, Andersson, & Pope, 1997; Nussbaum & Chaffin, 1996; Snook, 2001).

Biomechanics-based IPEs should be especially easy to establish given the close relationship between biomechanics and ergonomics.

Sports Sciences

Although we usually think of sports performance in terms of records at competitions (such as running speeds, distances thrown, or games won), the scientific assessment of the body's effort uses classic physiology and biomechanics procedures. However, there are two aspects to sports that differ from everyday tasks: In sports, we may attempt a one-time maximal performance instead of the submaximal efforts that usually characterize our daily work. The other particular feature is that usually selected healthy individuals, often especially gifted and motivated athletes, perform in sports. Yet, there are also persons with limited capabilities who participate in competitive sport, such as in the Special Olympics. Thus, much of the armory of sports sciences can be used to judge the performance of persons with disabilities and to establish IPEs.

Anthropometry

As discussed previously, body dimensions, particularly as they change with age, traditionally have been measured with conventional physical instruments (anthropometer, tape, caliper, and scale) between large numbers of loci on the body. Photographic measurements have not become widely accepted, but 3-D assessments (see above and Chapter 4) appear to be a promising technique of the future.

The following are traditionally taken measures:

- *heights*, such as stature, a straight-line, point-to-point vertical measurement
- *breadths*, such as hip breadth, a straight-line, point-to-point horizontal measurement running across the body or a segment
- *depths*, such as chest depth, a straight-line, point-to-point horizontal measurement running fore-and-aft the body
- *distances*, such as interpupillary distance, a straight-line, point-to-point measurement between landmarks on the body
- *circumferences*, such as waist circumference, a closed measurement that follows a body contour; hence, this measurement is not circular
- *reaches*, such as thumb-tip reach, a point-to-point measurement along the long axis of the arm or leg.

Chapter 4 presents more detailed explanations of anthropometric measurements. A large reservoir of anthropometric data is available (see the Appendix), but they generally stem from cross-sectional surveys of healthy adults. Only small and often incomplete data sets are available of older people, children, and pregnant women (see Chapters 5 through 8). New measurement techniques employ computerized procedures to measure and describe the human body.

Rehabilitation Engineering

The Available Motions Inventory (AMI) and the workability test (Dryden & Kemmerling 1990) are prominent among the classic, practical approaches in rehabilitation engineering to assess the specific abilities of persons with impairments.

The AMI consists of a set of miniworkstations arranged in various configurations around the participant. Among the more than 70 tests available are operations of switches, quantitative dial settings, rotations of hand cranks or knobs, assessment of the client's static and dynamic strengths, and performance of reaction-reach motions. The output of the AMI is a profile of the individual's abilities by "motion class" and a series of scores, which are based on a comparison of the individual's abilities with those of a group of able-bodied persons.

Workability denotes a series of workplace capability assessments. These were originally developed for clients with moderate disabilities using 21 tests with inexpensive materials, such as dice or marbles, to represent simple tasks common in industrial settings and offices. A second set of tests augmented the tasks, making them easier to apply and administer. The test results are compared with a benchmark standard scale.

To test the remaining abilities of persons with severe disabilities, specialized laboratory tests are applied. Some of these resemble standard biomechanical and psychological procedures, whereas others target specific activities ranging from propelling a wheelchair to performing occupational tasks (R. V. Smith & Leslie, 1990).

Industrial Engineering

Early in the 20th century, Lillian and Frank Gilbreth developed a method to decompose work activities into their elements (see, for instance, the account by Niebel &

Freivalds, 1999). They identified 17 basic elements of manipulation, called *therbligs*: transport empty, transport loaded, search, select, grasp, preposition, position, assemble, disassemble, release load, use, hold, inspect, avoidable delay, unavoidable delay, plan, and rest.

In the 1960s, the method of defining and measuring motion elements was refined so that task cycles, whole jobs, and complex industrial activities could be synthesized from activity elements. Based on similar procedures, a large variety of well-established elemental systems exists now throughout the industrial world; for example, British Standard 3138, MTM, REFA, and Work Factor. These systems are useful for two purposes: one is to predetermine the overall time for the execution of a task, and the other is to break down complex work into its basic elements. This facilitates modification of the task into a simpler one that is easier to do (Konz & Johnson, 2000).

This industrial engineering practice demonstrates that body motions and task activities can be successfully subdivided into a relatively small number of elemental activities. Conversely, complex tasks can be synthesized from these elements. This lets us measure an individual's ability to do such task elements in order to determine if that person can perform the overall task. It also allows for establishing in advance whether redesign of task or equipment would facilitate performance.

CONCLUSIONS DRAWN FROM THE REVIEW OF EXISTING TESTS

Various well-established tests exist in several scientific and applied disciplines. We can order them, starting with simple measurements of physical attributes and activities that require only minimal mental abilities, progressing to complex tasks with intensive use of cognition and intelligence:

- *body size measurements:* height, depth, breadth, circumference, reach, and weight
- *strength and mobility tests:* sets of biomechanical assessments, mostly related to the assessment of muscle capabilities such as in finger strength, grip strength, whole-body strength, and reach and motion capabilities
- *capacity tests:* measurements of health and fitness, usually via oxygen consumption or heart rate in reaction to experimental loadings
- *elemental motions:* a set of basic actions used to synthetically build up complex activities
- *sports and competitive tests:* assessments of such physical achievements as running speed, lifting strength, or games won
- *tests of available capabilities:* assessments of persons with disabilities to determine their remaining capabilities to perform specific tasks
- *manipulation and dexterity tests:* assessing performance in specifically designed laboratory tasks, such as the pegboard test
- *tests of everyday activities:* sets of common daily activities that comprise sets of elements, which may not be specifically identified
- *intelligence and cognitive tests:* assessment of mental and cognitive capabilities.

This brief listing demonstrates that sets of measurements and function tests exist, but at present these are largely limited to specific disciplines, are accepted and practiced only within those disciplines, and exhibit little cross-fertilization. This is unfortunate because there are obvious interrelationships. For example, tests of available capabilities used in rehabilitation contain elemental components similar to those in biomechanics, industrial engineering, health and fitness assessments, and in ADL and IADL.

POSSIBLE SOLUTIONS

Although desirable, it is obviously not feasible to combine *all* major features of these techniques in order to assess *all* testable capabilities of the human body and mind. That would require a large contingent of experts testing a given person over long periods and expensive and extensive preparation and instrumentation.

More practical is the approach of identifying specific task activities. Distinguishing ADL and IADL as components of daily life is a feasible example. But the currently used ADL and IADL comprise complex tasks that are clusters of (so far usually undefined) elemental capabilities. The ability to accomplish an ADL or IADL depends on the ability to perform the essential sets of elemental tasks (defined earlier as IPEs) that together comprise the total activity (Gloss & Wardle, 1982). If one element is impossible to perform, the whole activity cannot be done as planned.

A promising approach to developing a battery of ability tests is to perform the following steps:

- define IPEs drawn from various disciplines (as just discussed)
- combine them into sets of allied IPEs that are the essential building blocks of certain activities in daily life, occupational tasks, or leisure activities.

Once IPEs are defined, their performance by individuals can be measured objectively and quantitatively (as opposed to subjective, qualitative statements about competence, either by the individual or by a rater).

Combining allied IPEs into an ability test battery (ATB) would enable us to assess the performance capabilities of individuals and population groups at a given time (cross-sectional assessment) or the change in performance over time (longitudinal assessment).

Figure 2.1 displays the procedure of comparing capabilities and demands. It shows that the redesign of requirements, including the environment, is the fundamental solution to attaining a match between what we wish to do and what we can do.

SUMMARY

In order to match task requirements with individual capabilities, both must be known. This concept applies to all people, whether children, adults, pregnant women, older people, or persons with disabilities. It is critical for persons with limited capabilities. Once deficiencies are quantitatively known, ergonomic measures can be applied to compensate and overcome them.

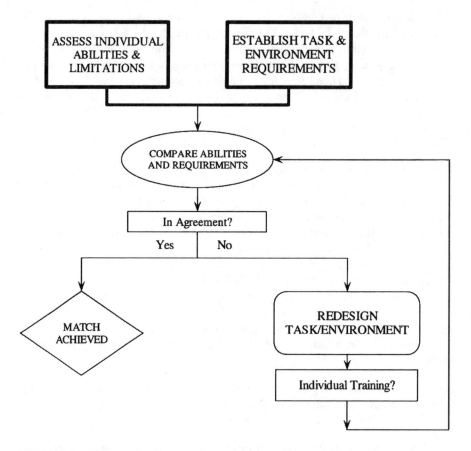

FIGURE 2.1 Achieving a match between abilities and task requirements.

Numerous measures of human capabilities and limitations are in use within various scientific disciplines. Many of these tests appear related and could be fruit-fully combined into sets, or test batteries, that measure the ability to perform tasks critical for daily life.

The ability to perform a task (or test) depends on the ability to execute its intrinsic performance elements, or IPEs. This understanding leads to the need to identify the critical IPEs, after which they can be measured objectively and quanti-tatively. Allied IPEs can be combined to form a comprehensive ability test battery that describes the match (or mismatch) between task demands and personal abilities.

3 Principles of Human Factors Engineering

OVERVIEW

Ergonomic system design is successful when it accommodates the person; otherwise, the person must struggle to fit the system. Matching task requirements or device dimensions to the user's characteristics ensures safety, ease, and proficiency.

Correct design relies on the recognition that humans are different from one another and that variations occur with age, pregnancy, health, or injury; the "average person" is a design phantom. This chapter contains a discussion of human factors data collections and their systematic use in design.

STRIVING FOR BETTER DESIGNS

The three major reasons to use human factors/ergonomics are associated with ethics, progress, and economics (Kroemer et al., 2001):

- The first motivation is the moral imperative to provide safety, ensure human health, generate comfort, and facilitate enjoyment.
- The second impetus is the quest to achieve progress in knowledge and technology. This means particularly to learn more about human desires, capabilities, and limitations, and consider these in the design of our environments, devices, and practices. Both goals contribute to improving the quality of life.
- The third reason is related to the economic advantages gained with reduced effort and cost in work systems with humans as doers, users, and beneficiaries. Stagnation in human factors engineering seems unacceptable because of the sentiment that things and conditions should be better in the future.

Most of the devices we use daily (such as can openers and automobiles), as well as complex systems (public transportation, for example), are primarily designed for normal users, ordinary use, and regular functioning. Yet, in reality, the vagaries of the environment can confound even the fittest person. Consider the extreme cases of a firefighter or soldier who is blinded by smoke or deafened by noise, whose mobility is impaired by terrain, whose dexterity is impaired by protective clothing, and who is cognitively impaired by high stress (Newell & Gregor, 2000). A passenger in a burning train is similarly baffled by such an unusually challenging environment, and so, to some extent, is a wheelchair-bound person stranded on an icy sidewalk by a sudden snowstorm or a senior citizen caught unawares in heavy rush-hour traffic.

The design task is probably becoming clear. Should we prepare all our systems and gadgets primarily for the most challenged user and the most complicated scenario and only then for normal conditions? This makes some sense, especially for complex systems and possibly dangerous devices; certainly, we should not exclusively design for fictitious regular use by normal adults. As detailed in Chapter 1, at least 40% of the global population would benefit from "extra-ordinary" ergonomics.

DESIGNING FOR SAFETY

Five major approaches allow for attaining safety through ergonomic design:

1. *Achieve fit* by proper sizing to ensure a good match between the object and its user. Examples are work clothing, car restraints, helmets, and office chairs. Ideally, the design should accommodate everybody, from the smallest to the biggest.
2. *Ensure reach* so that the human can access, use, and operate devices properly and effectively. Examples are door handles, controls on machinery, and handrails on stairs. Usually, small and weak persons present the critical reach-design criteria.
3. *Provide safe clearance* so that objects cannot hurt users or are not in the way. Examples are guards on circular saws, air bags in automobiles, escape passages, and legroom at a computer workstation. Large and tiny persons are the users who generally determine the safety design criteria.
4. *Avoid entrapment* of the body, for example, by railings on a child's bed or on a balcony, by an escape hatch, a repair opening in a structure, and pinch points in machinery. In many cases, a small body is critical, for instance, to determine the permissible spacing between rails, but the size of an escape hatch must accommodate a large person with bulky clothing.
5. *Provide exclusion*, for instance, by barriers and guards, so that dangerous spaces cannot be routinely or inadvertently accessed and dangerous actions cannot unintentionally be committed. Examples are placing an emergency button under a cover, allowing the gears of a car to be moved out of the "park" position only when the driver activates the brake at the same time. Here, no specific single group of users provides the design criteria; everybody must be considered.

DESIGNING FOR EASE OF USE

Many new designs of things and systems incorporate the "human factor" during the concept stage itself. Examples of well-thought-out designs are cell phones and computer mice, hip joint replacements, and spacecraft. They may not be perfect, but they are good.

The best strategy to design ergonomically is to do so right from the start, during the concept phase. The thinking of the concept team must be *use-centered,* which implies

- a goal — use for what?
- an instrument — use what?
- a process — use how?
- a user — who uses?

The goals of designing for easy (and safe and efficient) use, as discussed by Flach and Dominguez (1995), also apply to the improvement of existing systems. Examples are the infamous remote controls for home video players, bank automated teller machines (ATMs), pedestrian crossings at a busy road intersection, and such basic tools as a door handle and a light switch.

In spite of more than half a century of human factors engineering, there are still many products whose designs show little consideration of the user. One example is the old typewriter keyboard, transferred to today's computers (Kroemer 2001; K. H. E. Kroemer & A. D. Kroemer, 2001a). Replacement, retrofit, and redesign of improper devices make their usage possible or easier and avoid stress, strain, or injury. However, such redesign can be a costly and complex enterprise. An example is the provision of access to public buildings and transportation in the United States according to the Americans with Disabilities Act (ADA) of 1990. However, the effort did facilitate their use, not only for persons with disabilities but also for older people, for persons pushing baby strollers, and just about everybody else.

The need for enhancement or even retrofit may be obvious; for example, when a wheelchair user cannot climb a street curb. Deficiencies in performance (regarding quality or quantity) or, worse, injuries to people certainly raise a red flag. Other signs of an existing problem may be subtle: Why do so many older people avoid using certain ATMs? One common way to become aware of an existing design problem and pinpoint its specifics is careful interviewing of the users, who may even suggest a better solution.

Identifying the actual underlying problem is the most important step because failing to see the true reason invariably leads to selecting a candidate solution that misses the goal. Returning to the ATM example, why do so few older people use them? Even if the concept is new to them, why is it that they do not want to get used to it? Is it hard to read the display? Are the instructions difficult to understand? Is the placement of the controls inconvenient? Is the imposed fee the major deterrent? How about an inferred lack of security? Of course, the dislike may stem from a combination of several issues.

Take the computer workplace as another example (discussed by Kroemer et al., 2001): Ergonomically, one can make the layout of the room and the workstation better, improve the chair and other furniture (such as support for the display, support for the input device and the source document), enhance the lighting and the physical climate, reduce unwanted sound, and consider the need for privacy — but will any such measures really solve the problem? Or was the aversion of so many people to using a computer mostly a matter of perceived intellectual and skill demands?

DESIGNING FOR VARIABILITY

Taking human variability into account is the hallmark of ergonomics. Nobody stays the same during the course of one's life. All people are not the same. Externally,

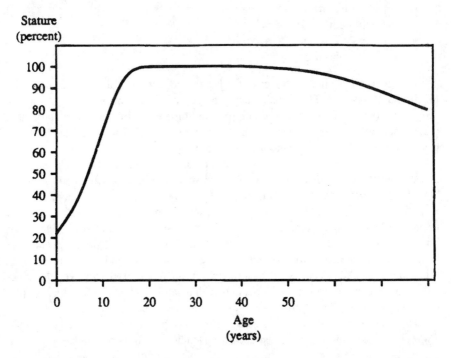

FIGURE 3.1 Approximate changes in stature with age. (From *Ergonomics — How to Design for Ease and Efficiency* (2nd ed.), by K. H. E. Kroemer, H. B. Kroemer, and K. E. Kroemer-Elbert, 2001, Upper Saddle River, NJ: Prentice Hall. Copyright 2001 by Prentice Hall. Reprinted with permission of Pearson Education, Upper Saddle River, NJ.)

they differ from one another in age, size, and strength. People are not made up of parts that are all average, or in any other single percentile value (see below); instead, every individual has different dimensions, physiological characteristics, and psychological traits that are unique to him or her.

The diversity of anthropometric, biomechanical, physiological, and psychological information can stem from four categories:

1. Within a person, body and segment size and strength and endurance change with age but also depend on nutrition, exercise, and health. Longitudinal studies, which observe an individual over years and decades, confirm such *intraindividual* changes. Most (but not all) variations within a person from infancy to old age follow the scheme shown in Figure 3.1. During childhood and adolescence, body dimensions such as stature and muscle strength increase rapidly. From the 20s into the 50s, little change occurs in general, but stature declines slightly. From the 60s on, many dimensions decline, whereas others, such as weight or bone circumference, often increase.

2. Individuals differ from one another in many respects. Data describing a population sample are usually collected in a cross-sectional study, in

which every person is measured at (about) the same moment in time, meaning that people of different age, health, and fitness are included in the sample set. The *interindividual* data found in most textbooks were gathered in cross-sectional studies. Table 1.1 in Chapter 1 and many other tables in this book (especially the anthropometric tables in the Appendix) are examples of such anthropometric and biomechanical data compilations.

3. There is some factual and much anecdotal evidence that European, North American, and Asian people nowadays are on average larger (although possibly neither stronger nor more able to do demanding work) than their ancestors. Also, life expectancy has much increased, so the proportion of older people in the workforce and in the overall population is much larger now than just a few decades earlier. The probable reasons for this secular development are improved nutrition, hygiene, and health care, which have allowed people to achieve more of their genetically determined potential than was possible generations ago.

4. Finally, there is the possibility of variance in measurements owing to differences in selecting population samples, using measurement techniques, storing the measured numbers, and applying statistical methods to the raw data.

Using Statistics

The basic and simple statistical descriptors *mean* (*m;* often called *average*), *standard deviation* (*S*), and *sample size* (*n*) fully describe normal distributions of data. Fortunately, many ergonomic data, and anthropometric statistics in particular, appear as reasonably normal (Gaussian) distributions, meaning that they follow the bell-shaped pattern (symmetrical on both sides of the average). However, there are some important exceptions; for example, in muscle strength and other physiological and psychological information. Data on special populations and subject samples, discussed in this book, may not show normal distributions.

One easy way to check on data diversity of a normal distribution is to divide the standard deviation in question by its average. This yields the *coefficient of variation*, CV. In most cross-sectional studies of adult body dimensions, the CV is around 5%, but in many strength data, it is up to 25% or even larger.

In wanting to determine how one variable *y* (for example, arm length) changes with another variable *x* (say, stature), one often assumes that *y* varies directly with *x*. (However, few people ever check the correctness of that assumption of a linear regression.) The correlation coefficient *r* describes such a linear relationship. The *coefficient of determination* is the square of the correlation coefficient: R^2 indicates the proportion of variation in the dependent variable *y*.

The many textbooks on statistics contain more mathematically precise definitions.

Phantoms, Ghosts, and the "Average Person" Manikin

Actually, relationships among body dimensions vary enormously. Table 3.1 shows this for U.S. Army soldiers. The correlations *r* between their body size measurements digress greatly, from negative to positive; they can range from 0 (meaning no

TABLE 3.1
Correlations Between Anthropometric Data on U.S. Soldiers. (Values for women are listed above the diagonal, below for men. Values larger than 0.7 carry an asterisk.)

	1 Age	2 W	3 St	4OFR	5 WH	6 CH	7 SH	8 PH	9 SC	10CC
1 Age [302]		.219	.041	.017	.044	-.055	.066	-.074	.155	.193
2 Weight [125]	.195		.529	.493	.491	.370	.422	.242	.845*	.806*
3 Stature [100]	-.021	.546		.928*	.848*	.840*	.755*	.808*	.377	.222
4 Overhead fingertip reach[84]	-.013	.525	.937*		.704*	.905*	.554	.868*	.384	.199
5 Wrist height, standing [128]	.028	.527	.856*	.749*		.625	.754*	.587	.300	.255
6 Crotch height [39]	.090	.351	.852*	.890*	.673		.330	.915*	.267	.093
7 Sitting height [94]	.026	.447	.741*	.578	.692	.347		.343	.285	.202
8 Popliteal height, sitting [87]	-.094	.341	.852*	.883*	.673	.924*	.383		.188	.023
9 Shoulder circumference [91]	.122	.861*	.399	.413	.334	.250	.326	.256		.808*
10 Chest circumference [34]	.279	.873*	.312	.308	.357	.135	.287	.137	.859*	
11 Waist circumference [115]	.364	.849*	.276	.251	.343	.060	.298	.074	.703*	.839*
12 Buttock circumference [24]	.190	.935*	.401	.380	.412	.204	.373	.191	.781*	.815*
13 Span [99]	-.016	.497	.815*	.908*	.535	.840*	.398	.844*	.445	.281
14 Biacromial breadth [11]	.034	.496	.487	.506	.295	.370	.407	.394	.633	.419
15 Hip breadth, standing [66]	.209	.831*	.453	.416	.457	.239	.464	.224	.672	.727
16 Head circumference [62]	.125	.508	.342	.312	.302	.224	.303	.240	.433	.421
17 Head length [63]	-.002	.371	.346	.315	.295	.260	.302	.268	.295	.271
18 Head breadth [61]	.198	.320	.114	.098	.112	.034	.128	.035	.303	.311
19 Hand length [60]	.032	.453	.650	.724*	.464	.676	.300	.679	.372	.242
20 Foot length [52]	.012	.512	.700	.734*	.537	.687	.383	.697	.409	.299

	11 WC	12 BC	13 Sp	14 BB	15 HiB	16 HC	17 HeL	18 HeB	19 HaL	20 FL
1 Age [302]	.299	.258	.011	.025	.283	.073	.027	.044	.044	.026
2 Weight [125]	.767*	.897*	.438	.440	.778*	.428	.329	.420	.430	.493
3 Stature [100]	.167	.361	.787*	.505	.372	.348	.354	.124	.637	.673
4 Overhead fingertip reach[84]	.132	.313	.907*	.535	.294	.337	.345	.095	.737	.732*
5 Wrist height, standing [128]	.217	.363	.453	.303	.397	.250	.261	.403	.403	.468
6 Crotch height [39]	.061	.185	.870*	.418	.146	.287	.302	.043	.706*	.703*
7 Sitting height [94]	.142	.351	.336	.384	.438	.246	.255	.159	.256	.330
8 Popliteal height, sitting [87]	-.031	.063	.840*	.420	.051	.241	.271	.020	.685	.671
9 Shoulder circumference [91]	.697	.726*	.395	.574	.601	.353	.264	.261	.355	.379
10 Chest circumference [34]	.781*	.707*	.167	.304	.603	.393	.191	.246	.186	.288
11 Waist circumference [115]		.738*	.109	.214	.673	.223	.117	.229	.127	.170
12 Buttock circumference [24]	.859*		.258	.327	.915*	.313	.226	.220	.258	.323
13 Span [99]	.201	.352		.565	.203	.345	.338	.083	.827*	.775*
14 Biacromial breadth [11]	.311	.411	.575		.294	.287	.259	.152	.441	.456
15 Hip breadth, standing [66]	.799	.902*	.355	.404		.232	.160	.196	.180	.250
16 Head circumference [62]	.376	.427	.320	.301	.364		.824*	.497	.342	.360
17 Head length [63]	.222	.301	.304	.235	.259	.820*		.131	.337	.339
18 Head breadth [61]	.277	.268	.131	.180	.235	.541	.120		.082	.113
19 Hand length [60]	.166	.320	.810*	.433	.298	.330	.306	.137		.825*
20 Foot length [52]	.220	.390	.766*	.445	.377	.333	.304	.161	.806*	

Note. All pairs of data that correlate above .700 appear both in the male and female groups; the correlations below .300 are similar for both genders.

Data, including numbers in brackets, were taken from the *1988 Anthropometric Survey of U.S. Army Personnel* (Technical Reports 90/031 through 036), by J. Cheverud, C. C. Gordon, R. A. Walker, C. Jacquish, L. Kohn, A. Moore, and N. Yamashita,1990, Natick, MA: U.S. Army Natick Research, Development, and Engineering Center.

relationship) to 1 (indicating a perfect relationship). The attempt to express all body characteristics in terms of one basic denominator is futile. Yet, in the past, too many designers used a scheme that, deceptively, expressed heights, breadths, depths, and segment lengths as a percentage of stature. For instance, average hip breadth was flatly stated by one so-called authority to be 19.1% of standing height, which makes no sense at all because hip breadth varies widely between males and females as groups and among individuals. (Besides, it is hard to think of any object that could be properly designed for a fixed average hip breadth.)

Another statistical folly is to think of a human body as consisting of parts that are all of the same percentage value. Such persons do not exist, and anthropometric statisticians have proven that concept irrational. Yet, in spite of the obvious fallacy of the single-percentile model, various phantom templates have appeared inexplicably in which body segments are all of the same percentile value. The ghost most often encountered is the 50th percentile pattern, the proverbial average person. Other ghostly figures have, for example, all 5th or 95th percentile values. Of course, designs for these figments of the imagination hardly fit actual users.

Surprisingly, some engineers — even physiologists and physicians — are still willing to express body attributes in terms of stature (probably because height is easily measured), even when a proper logical or statistical correlation is lacking. For example, body weight is often expressed in relation to standing height, in spite of the fact that the correlation between these two variables is only around 0.5, at best. Check Table 3.1 for some actual data, keeping in mind that most active soldiers are fairly fit persons. In the general population, the correlation between stature and body weight is certainly much lower.

PROPER DESIGN PROCEDURES

The basic task is, *first,* to *identify the critical human attribute* (often several attributes) and, *second, to accommodate it* (them) by careful design (HFES 300 Committee, 2004; Kroemer et al., 2001; Robinette, 1998). Having established the crucial characteristic, we must quantify it so we can ascertain its critical values, usually in terms of lower and upper limits.

Here are a few examples:

- Small children may put their heads between balusters of railings and then become stuck. After measuring children's head sizes, the architect recognizes that the breadth of the head is critical. Therefore, the opening between neighboring pickets must be smaller than the narrowest skull observed.
- Many older persons find it especially difficult to step up, for example when they want to board a public bus. After measuring on an appropriate sample of bus users the ability to climb steps of different heights, the designer determines the height that at least two thirds of them can climb with acceptable effort. One solution that avoids steps altogether is to make

buses "kneel" at the boarding stop so that the surfaces of curb and bus floor become approximately level.

- In an assembly plant, stamping presses used to join parts may injure the hands of persons with exceptionally long and slender fingers. Extending the safety cage around the pinch point farther out than the operator with the longest hands could reach in solves that problem.

SINGLE CUTOFF

The three foregoing examples show how the ergonomist determines a single anthropometric value (the narrowest head, tallest step, and longest hand) that decides a design dimension, such as the minimal baluster spacing, the desired step-up height, and the necessary safety clearance. These solutions utilize the simple principle of *"max or min"* design.

"Max *or* Min" Design

This involves applying either of the following principles:

- design for the maximal value — for the big, the burly: the upper limit

OR

- design for a minimal value — for the small, the feeble: the lower limit

The common way to determine the critical value is through statistical procedures (illustrated below the section How to Determine Critical Design Values) applied to relevant compiled data sets. Occasionally, we must ascertain such single cutoff values by specifically measuring a carefully selected sample of people of concern.

DUAL CUTOFF

"Min *and* Max" Design

In many applications, lower and upper values *both* determine the appropriate design. They establish adjustment limits, such as the shortest and tallest heights between which we can set an office chair to accommodate the different lengths of the lower legs of the user. Other examples for the minimal and maximal design practice are the clothing tariffs that determine that slacks, shirts, gloves, and hats are cut so that each size group fits the body dimensions of a certain segment of the wearers.

Setting the upper and the lower design limits is common in industry. Many U.S. military procurement requirements stated that the range from 5th to 95th percentile of soldiers must be accommodated by equipment. This is a laudable example of setting minimal and maximal limits, but, unfortunately, the military documents commonly lack a definition of the crucial characteristics for which these limits shall be fixed. Clearly, one cannot assume that soldiers belong exclusively to either the 5th or the 95th percentile group. Another example of setting minimal and maximal limits is in establishing clothing sizes (Laing, Holland, Wilson, & Niven, 1999; McConville, 1978).

Use of Average Values

After attending a human factors short course, the head of the operations department in an oil refinery became concerned that not all personnel might be strong enough to close emergency shutoff valves in the pipelines. He ordered that the employees be tested and then all valve controls altered, if necessary, so that they would not require more than average strength for closing. He submitted an article describing this human-engineered solution to a professional journal, and the editor promptly rejected it.

It is hard to imagine even one case in which something could be properly designed to fit the average of a data set. Why? Because whatever is designed for a mean value is too little for half the users and too much for all the others, as logic suggests and experience confirms. Employing the average as a design guide invariably cuts off about half the users — as did, unthinkingly, the oil refinery boss.

Although the mean of a distribution is very unlikely to serve as a *criterion* for design, it is a useful and necessary *tool* for statistical purposes, as discussed in the following text.

How to Determine Critical Design Values

We can specifically measure persons who must be accommodated in order to determine critical percentile points for design dimensions; for example, single design cutoffs or min-max limits. This is usually done individually for each person who has a disability that is to be overcome by a particular design. How such measurements should be taken depends on the individual case, but publications by, for example, Chaffin, Andersson, and Martin (1999), HFES 300 Committee (2004), Kroemer (1997a), Kroemer (2004), and Roebuck (1995) contain guidelines for anthropometric and biomechanical assessments.

It is time consuming and expensive to take such measurements for a large sample of people or even whole populations. That may explain why, for example, in the United States children have not been measured comprehensively in terms of anthropometry, strength, and biomechanical properties since the 1970s (see Chapter 8). Such measurements taken on large samples provide data that can provide useful tables of statistics, such as listed in the Appendix. If the designer is lucky, such tables contain exactly the data points needed, such as 5th or 95th percentile values. Otherwise, the designer must calculate or estimate the desired numbers, as explained below.

There is another way to determine given percentile points without any formal calculations if the data are available graphically as a curve. In this case, one simply measures, counts, or estimates the desired percentile values from the graph. This procedure works whether the distribution is normal, skewed, binomial, or in any other form. For example, the U.S. Centers for Disease Control regularly updates growth charts of U.S. children, which are easy to download and use. To determine a data point on one of the curves provided, we read the value at the age intersection of the graph (see Chapter 8). If the required percentile does not fall on any one existing curve, we can interpolate between the curves.

STATISTICAL PROCEDURES

If a set of data has a normal (Gaussian) distribution, we can fully describe it using just two descriptors. The *mean* (*m;* same as *average*) identifies the central point of the distribution; 50% of the data lie below and the other 50% above the mean, which is therefore called the *50th percentile* (*p50*). The other main descriptor is the *standard deviation* (*S*), which specifies the spread of the data about the mean. The numerical value of the standard deviation is larger when the data are widely scattered than when they cluster close to the mean.

Using these expressions, the location of any (percentile) point in the distribution can be simply calculated from the mean and standard deviation. To calculate a percentile value, simply multiply the standard deviation S by a factor k selected from Table 3.2. Then add the product to the mean m:

$$p = m + (k \times S)$$

If the desired percentile is above the 50th percentile, the factor k has a positive sign and you add the product $k \times S$ to the mean m. If the percentile is below average, k is negative, and accordingly you subtract the product.

Examples

1st percentile is at $m - 2.33S$	with $k = -2.33$ (see Table 3.2)
5th percentile is at $m - 1.64S$	with $k = -1.64$
10th percentile is at $m - 1.28S$	with $k = -1.28$
50th percentile is at m	because $k = 0$
60th percentile is at $m + 1.28S$	with $k = 1.28$
95th percentile is at $m + 1.64S$	with $k = 1.64$

Determining a Single Percentile as Cutoff Point (Max *or* Min)

This procedure has the following steps:

1. Select the desired percentile value p.
2. Determine the associated k value from Table 3.2.
3. Calculate the p value from $p = m + (k \times S)$. (Note that k, and hence the product, may be negative.)

Determining the Upper and Lower Limits of a Range (Min *and* Max)

This procedure has the following steps:

1. Select upper percentile p_{max}.
   Find related k_{max} value in Table 3.2.
   Calculate upper percentile value $p_{max} = m + (k_{max} \times S)$.

TABLE 3.2
Percentile Values With Their k Factors

Below Mean				Above Mean			
Percentile	Factor k	Percentile	Factor k	Percentile	Factor k	Percentile	Factor k
0.001	−4.25	**25**	**−0.67**	**50**	**0**	76	0.71
0.01	−3.72	26	−0.64	51	0.03	77	0.74
0.1	−3.09	27	−0.61	52	0.05	78	0.77
0.5	−2.58	28	−0.58	53	0.08	79	0.81
1	−2.33	29	−0.55	54	0.10	**80**	**0.84**
2	−2.05	**30**	**−0.52**	**55**	**0.13**	81	0.88
2.5	−1.96	31	−0.50	56	0.15	82	0.92
3	−1.88	32	−0.47	57	0.18	83	0.95
4	−1.75	33	−0.44	58	0.20	84	0.99
5	**−1.64**	34	−0.41	59	0.23	**85**	**1.04**
6	−1.55	**35**	**−0.39**	**60**	**0.25**	86	1.08
7	−1.48	36	−0.36	61	0.28	87	1.13
8	−1.41	37	−0.33	62	0.31	88	1.18
9	−1.34	38	−0.31	63	0.33	89	1.23
10	**−1.28**	39	−0.28	64	0.36	**90**	**1.28**
11	−1.23	**40**	**−0.25**	**65**	**0.39**	91	1.34
12	−1.18	41	−0.23	66	0.41	92	1.41
13	−1.13	42	−0.20	67	0.44	93	1.48
14	−1.08	43	−0.18	68	0.47	94	1.55
15	**−1.04**	44	−0.15	69	0.50	**95**	**1.64**
16	−0.99	**45**	**−0.13**	**70**	**0.52**	96	1.75
17	−0.95	46	−0.10	71	0.55	97	1.88
18	−0.92	47	−0.08	72	0.58	98	2.05
19	−0.88	48	−0.05	73	0.61	99	2.33
20	**−0.84**	49	−0.03	74	0.64	99.5	2.58
21	−0.81	**50**	**0**	**75**	**0.67**	99.9	3.09
22	−0.77					99.99	3.72
23	−0.74					99.999	4.26
24	−0.71						

Note: Any percentile value p can be calculated from the mean m and the standard deviation S (normal distribution assumed) by $p = m + (k \times S)$.

2. Select lower percentile $p_{min.}$
    Find related k_{min} value in Table 3.2.
    Calculate lower percentile value $p_{min} = m + (k_{min} \times S)$.
3. Determine range $R = |p_{max}| − |p_{min}|$.

(Note that the two percentile values [p_{max} and p_{min}] need not be at the same distance from the m, i.e., the range does not have to be symmetrical about the mean, and one percentile or both percentiles may be located either above or below the mean.)

Combining Anthropometric Data Sets

Occasionally we must combine anthropometric values by adding or subtracting them; for example, total fingertip reach length is the sum of the lengths of upper arm, forearm, and hand.

Combining measures, such as leg length and torso (with head) length, generates a new distribution, stature. For combining, one must consider the *covariation COV* between the two measures; for example, usually (but not always) a taller torso is associated with a longer leg. This is mathematically described by the correlation coefficient r between the two data sets, x and y:

$$COV(x,y) = r_{x,y}\, S_x\, S_y.$$

As usual, we calculate the *sum of the mean values* of the x and y distributions from

$$m_z = m_x + m_y.$$

The estimated standard deviation S_z of the mean of the new variable z is

$$S_z = (S_x^2 + S_y^2 + 2r\, S_x\, S_y)^{1/2}.$$

Similarly, the *difference between two mean values* is

$$m_z = m_x - m_y.$$

and its standard deviation

$$S_z = (S_x^2 + S_y^2 - 2r\, S_x\, S_y)^{1/2}.$$

The following example illustrates the application of this procedure to answer the question:

What is the p95 shoulder-to-fingertip length?

Assume the average forearm (FA) length (with the hand) to be 442.9 mm with a standard deviation of 23.4 mm. Also assume an average upper arm (UA) length of 335.8 mm with a standard deviation of 17.4 mm.

Produre: You cannot add the two p95 lengths because this would disregard their covariance; instead, calculate the sum of the mean values first:

$$m = m_{FA} + m_{UA} = 442.9 + 335.8 = 778.7 \text{ mm}$$

Next, calculate the standard deviation using an assumed coefficient of correlation of 0.4:

$$S = [23.4^2 + 17.4^2 + (2 \times 0.4 \times 23.4 \times 17.4)]^{1/2} \text{ mm} = 34.6 \text{ mm}$$

Now you can determine the p95 total arm length (AL) using k = 1.64 (from Table 3.2):

$$AL_{95} = 778.7 \text{ mm} + (1.64 \times 34.6 \text{ mm}) = 835.4 \text{ mm}$$

The total p95Arm Length AL, shoulder-to-fingertip, is 835 mm.

HOW TO GET MISSING DATA

We may have to design a new product for users about whom we do not have precise information on crucial design values. One example of such a task is designing protective sports gear for American teenagers. Little formal knowledge about their body sizes, strength capabilities, and other biomechanical descriptors exists in the open literature, but these measures are needed for proper design.

For precise and comprehensive information, more is necessary than just casually measuring a few acquaintances. Two avenues are open: One is to conduct a formal anthropometric survey, but this is a major enterprise and best done by qualified anthropometrists. The other option is to deduce from existing data what we need to know. There are several approaches to estimating missing data (Kalton & Anderson, 1989).

Estimation by *ratio scaling* (in 1996, Pheasant used it extensively) is one technique to estimate data from known body dimensions. It relies on the assumption that although people vary greatly in size, they are likely to be similar in proportions. (Note that this assumption is debatable.) This premise probably holds true for body components that are related in size to each other, as shown in Table 3.1. Among adults, many body lengths are highly correlated with each other; also, groups of body breadths are related, as are circumferences as a group. However, not all body lengths, breadths, or circumferences are highly correlated with each other, and certainly many lengths are not highly correlated with widths, breadths, or circumferences. Thus, one has to be very careful in deriving one set of data from another. This is particularly true in the case of teenagers, whose body proportions change drastically from late childhood to early adulthood.

A good rule for ratio scaling is to use only pairs of data that are related to each other with a coefficient of correlation of at least 0.7. (This ensures that the variability of the derived information depends by about 50% on the variability of the predictor: Squaring the correlation coefficient 0.7 results in a value almost equal to 0.5.) For sets of highly correlated data, ratio scaling can be an appropriate means of estimating missing but required design guide numbers, as the HFES 300 Committee (2004), Kroemer (1999b), and Roebuck (1995) have discussed in some detail.

Another way of estimating the relationships between body dimensions is through *regression equations*. Most commonly used regression equations are bivariate in nature, and the assumption is that the two variables relate linearly with each other. (This postulated linear relationship is seldom explicitly confirmed.) Then, the general form is

$$y = a + (b \times x)$$

where *x* is the known mean value and *y* the predicted mean. The constants *a* (the "intercept") and *b* (the "slope") must be determined (known) for the data set of interest. An example of this procedure is the estimation of body dimensions of American soldiers by Cheverud et al. (1990).

When trying to predict the mean value of *y* (for any value of *x*) by using the regression equation shown above, we must remember that the actual values of *y* are scattered about the mean in a normal (Gaussian) probability distribution. The standard error SE of the estimate depends on the correlation *r* between *x* and *y* (and the standard deviation of *y*, S_y) according to the following equation:

$$SE_y = (1 - r^2)^{1/2}$$

Roebuck (1995) discussed this concept in detail, including its extension to the development of multivariate regression equations, principal component analyses, and boundary description analyses.

We must realize here that these statistical applications categorically assume representative (usually large) sample sizes with normal (Gaussian) distributions, which is often true for data on normal populations but does not usually apply to samples of "extra-ordinary" persons. In the latter case, great caution is in order when applying ordinary statistics.

DESIGNING FOR BODY STRENGTH

Designing for human physical strength (in terms of the force or torque that a person can exert) is a frequent task for the human factors engineer. It involves a number of decisions:

- Is strength use mostly static or dynamic? If it is static, we may employ information about isometric strength capabilities. If it is dynamic, other considerations apply in addition, concerning, for example, physical (circulatory, respiratory, and metabolic) endurance capabilities of the operator and prevailing environmental conditions. Physiological and ergonomics texts (such as Astrand & Rodahl, 1986; Kroemer, 1997a; Kroemer, 1997b; Kroemer et al., 2001; Kumar, 2004, Schmidt & Thews, 1989; D. A. Winter, 1990) provide such information. As a rule, strength exerted in motion is less than that measured in static positions located on the path of motion.
- Is the exertion in slow or fast movement? Most body strength data available concern static (isometric) exertions. This information gives fair guidance also for slow motions but not for fast actions (Kroemer, 2004).
- Is the exertion by hand or by foot, or does it involve the whole body? For each of these situations, the ergonomics literature provides specific design information (Chengular, Rodgers, & Bernard, 2003; Karwowski & Marras, 1999; Kumar, 2004).
- Is a maximal or a minimal strength exertion the critical design factor? *Maximal* user output usually determines the structural strength of the

object in order that the strongest operator may not break a handle or a pedal. The design value is set, with a safety margin, above the highest perceivable application of strength. *Minimal* user output is that exertion expected from the weakest operator that still yields the desired result, so that a door handle or brake pedal can be successfully operated or a heavy object moved. (See the previous discussion for more on upper and lower limits.)

DESIGNING THE "SIGNAL LOOP"

As mentioned in Chapters 1 and 2, the traditional explanation of the functioning of the human mind relies on a construct made up of sequences (or stages) that follow a signal input. The idea is that sensation leads to perception, then to decoding, followed by decision, all resulting in an appropriate response. This concept appears overly simple to most neurophysiologists, psychologists of various specializations, and ecologists, all of whom have generated a great diversity of special models. These include multiple parallel feedforward and feedback paths, simultaneous and sequential impulses, and complex processing, which refine the traditional concept (Arswell & Stephens, 2001; Vicente, 2002).

Figure 3.2 shows a traditional model of the human as a linear processor of signals. A sensor detects information (in some perceptible form of energy) and sends impulses along the *afferent* (sensory) pathways of the peripheral nervous system (PNS) to the central nervous system (CNS). The CNS processes the signals and chooses an action (including the "no action" choice). Appropriate feedforward impulses are generated and transmitted along the *efferent* (motor) pathways of the PNS to the *effector* such as voice, hand, and foot. A feedback loop serves to compare the output of the system with the desired input. If a disparity between actual system performance and the intended performance exists at the comparator, this signal is used to modify system performance.

One of the major traditional tasks of the ergonomist is to check the nature of the external signal and determine whether humans can perceive it. If so, it is often still necessary to modify the incoming signal to facilitate its perception. Examples are modifying the color of a light, modulating the sound of a siren, and adjusting the qualities of a speaker system. If the signal is not within the "bandwidth" of the human senses, the nature of the signal must be changed altogether. Examples of such tasks are making the presence of radiating energy known by generating a buzzing or flashing signal, or projecting an image ("head-up" display) onto the windshield of an airplane or car. In many cases, tweaking the input also includes correcting conditions that affect the signal itself or its ability to pierce the clutter of the environment. The illumination levels of computer offices, for example, must be kept rather low to avoid glare on the computer screens. The dim lighting, however, can make reading of paper documents difficult, especially for older persons. This can require the use of task lights, but these need to be placed carefully so as not to generate glare on the computer monitors (K. H. E. Kroemer & A. D. Kroemer, 2001a). Providing proper input signals, as shown in Figure 3.3, can be a daunting ergonomic and engineering assignment.

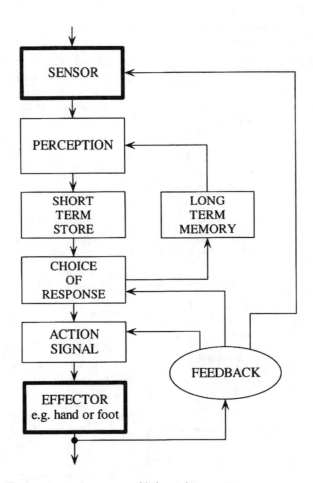

FIGURE 3.2 The human as a processor of information.

 On the output side, the actions of the human effector (usually hand or foot) may directly control the machine, such as a bicycle or automobile. Certain conditions may generate a need for yet another transducer; for example, movement of a steering wheel by the human hand may require auxiliary energy (power steering). Figure 3.4 portrays the model. Recognizing the need for a transducer and providing information for its suitable design is again a primary task for the ergonomist or human-factors engineer.

The human engineering tasks of providing proper output controls to operate a machine or process are often demanding. Steering a bicycle by turning the handle bar seems simple, but it takes a novice a long time to get the feel for proper control actions, and changing to a different handle on another bicycle requires new learning. To avoid overloading the human operator, auxiliary power in the control module is often needed. An example of this is the steering of cars and trucks, where, in addition to power-assisted steering, the relationship between human effector movement of the control and the resulting vehicle response is made variable because the ratio

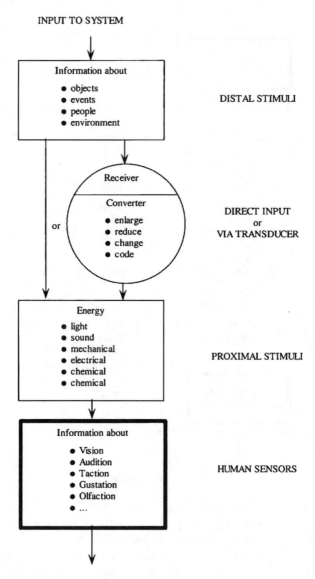

FIGURE 3.3 Transforming distal signals into proximal stimuli that can serve as inputs to the human processor.

between steering wheel turn and the heading angle of the steered wheels depends on actual speed and current steered angle.

In complex cases, the effector output does not result immediately in a perceivable response of a system (an example is changing the rudder setting of a large ship), and the actual system behavior may depend on various existing conditions, such as the strength and direction of a current in which a ship or airplane maneuvers. In such cases, some kind of prediction logic, incorporated into the system, tells the

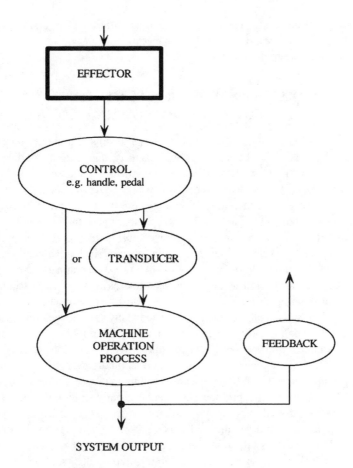

FIGURE 3.4 Transforming human effector outputs.

human operator about the future response of the system. This predictive information, in turn, allows the operator to modify the effective control.

The ergonomics engineer must provide the human with suitable transducers on both the input and the output sides of the system. Yet, the same external stimulus may trigger various — even unpredictable — responses in different persons. What is an adequate solution for the prototypical ordinary adult may not be appropriate for a child, an older individual, and a person with a physical disability. Simple reaction time (see Chapters 1 and 2) changes little with age from about 15 to 60 years but is substantially slower in children. Reaction and response times may slow moderately in some persons but severely in others as they grow old. This indicates likely deterioration in peripheral and central functions with aging, often accompanied by reductions in physiological and biomechanical functions. Pregnancy brings about difficulties and handicaps that are usually temporary, whereas many persons with disabilities have specific dysfunctions that are permanent.

With the current understanding of the human mind, it is still more visionary than realistic to expect that thought can control devices. Yet, some steps in that direction have been made; for example, in rehabilitation engineering, one may electrically stimulate paralyzed muscles for feedforward or use stimulation of skin sensors for feedback control of prostheses.

DESIGNING FOR GROUPS OF PEOPLE
AND FOR INDIVIDUALS

We receive information about the outside, external events and about other persons by the senses of vision, audition, olfaction, gustation, and taction. As discussed in Chapters 1 and 2, we know well how the eyes function, how to measure these functions, and how age, illness, or injury can affect them. Similarly, we understand the functions of the hearing sense well, including its changes due to aging and noise. Although the sensations related to taction (touch, pain, temperature, and electricity), as well as smell and taste, have been researched for centuries, they are still not carefully enough defined and measured to be reliably employed by the human factors engineer.

In general, however, these senses work in the same manner from person to person. Thus, we can design technical systems and assistive devices so they meet the seeing and hearing needs of most persons. However, as regards the senses of taction, olfaction, and gustation, some general and many specific uncertainties prevail even for regular adults, and changes experienced from childhood to advanced age or those associated with illness or injury remain difficult to define and measure. This uncertainty explains imprecise, tentative, and occasionally even contradictory statements in the literature. Much basic research on human senses still needs to be done (or redone), and the precise problems of special groups or individuals necessitate specific assessments so that proper ergonomic designs can meet their needs.

An example of such assessment of needs and a derived ergonomic solution is that of signaling when it is safe for a pedestrian to cross a busy road at an intersection. The normal solution is to provide a traffic light setup, with green meaning *go* and red meaning *stay*. For people who cannot see or those who cannot distinguish between red and green, additional acoustic indicators would be helpful; for example, a chirping sound for *go* and a cuckooing sound for *stay*. This would help even regular persons, as well as children and older people. However, it does not help people who are blind and deaf, and naturally it does nothing for those who have difficulties ambulating.

Basic functions in physiology (the interacting systems of metabolism, circulation, and respiration) and biomechanics (such as speed and forcefulness of motions) essentially perform in the same way from early childhood to advanced age. However, their results and outputs (as measured by force, work, or endurance) are variable, not only among age groups but also, for example, during the course of pregnancy or among persons with disabilities. Likewise, the thermoregulation of the body performs differently in children, pregnant women, and older people compared to healthy adults. Furthermore, different climatic regions of the Earth are associated with different physical and behavioral traits of the people living there — yet, little systematic knowledge is available for the ergonomist.

Table 3.3 contains an overview of trends of changes in the functions discussed in the preceding text that are of main concern to the ergonomist. However, these

TABLE 3.3
Trends of Changes in Functions Compared With Normative Adults

	Energetic Capabilities (metabolism, circulation, respiration)	Biomechanical Capabilities (strength, power mobility, endurance)	Sensations, Perception Decision Making, Controlling Actions	Capabilities to Endure and Function in Heat and Cold	Stress Tolerance
Pregnancy	Reduced, diminishes further with run of pregnancy	Reduced, diminishes further with run of pregnancy	No change except in taste and smell	Reduced, diminishes further with run of pregnancy	Reduced, diminishes further with run of pregnancy
Childhood	Much smaller but increases	Much smaller but increases	Not systematically known	Much smaller but increases	Not systematically known
Aging	Reduced, diminishes further with aging	Reduced, diminishes further with aging	Reduced, diminishes further with aging	Reduced, diminishes further with aging	Depends on individual and circumstances
With permanent disability	No general statements possible, depends on the individual				

changes can be different for individuals in their temporal development, magnitude, and importance. Therefore, it is often necessary to assess individuals' capabilities or limitations with respect to certain tasks or equipment in personalized tests. The biomechanical, gerontological, orthopedic, and rehabilitation literatures provide examples and guidance.

Nearly all our everyday systems are designed for persons who respond correctly and rapidly to signals coming from their surroundings. It can be complicated for a child, a pregnant woman, or a handicapped person to function in our artificial world designed for the healthy and fast-responding adult. When the pedestrian light over the crosswalk of a busy street shows green, a backache slowing our gait can make it difficult to cross in the time allotted by the signal. As drivers, we find it difficult to quickly adjust our responses when, unexpectedly, the power booster for brakes and steering in our car fails; this may be a catastrophic event for the proverbial little old man driving to church on Sunday morning.

SUMMARY

Designing for variability can require measurement of individual capabilities (or the lack thereof) of the one person whom we want to help overcome his or her limitations. This usually demands employing special laboratory equipment and measuring procedures, such as discussed in Chapter 2. In most cases, however, the human factors engineer will design tasks and hardware for groups of people, in many cases for large population subsamples. "Design for many" instead of for individuals commonly incorporates selection and use of statistical data on critical design-related traits such as body size, physical strength, mobility, and abilities to see and hear. The lower and upper design parameters chosen determine the usability, ease, and safety of use of the process or product.

Although the statistical average is a descriptor of the central properties of a data distribution, people are not made up of parts that are all average, nor do they appear in any other given single percentile configuration. Instead, every individual has unique dimensions, physiological characteristics, and psychological traits that rank in different percentile ranges. Taking human variability into account is the hallmark of ergonomics.

4 Design for Movement: With Special Solutions for the Very Small and Big, for Those With Lower Back Problems, and for Bedridden Persons

OVERVIEW

For convenience, we traditionally measure human body size and static body strength while the person holds still in a defined body posture. But in real life, we work, walk, and sit while continually changing our body's configuration. The human body is intended for motion, and therefore we must design our equipment and tasks for movement as well.

People come in all sizes and patterns, many different from the norm. At times, they are not healthy or need to recover from an injury, in a hospital or at home. Low back problems are rampant. There are ergonomic solutions to make life a bit easier in all such conditions.

This chapter first discusses how to design for the moving body, then how to measure body size, and finally what specific measurements are required for equipment so that it can be designed to fit persons of other than average size, who have special needs for ergonomic help. We can all use some special attention, if not every day, then at least occasionally.

DESIGN FOR MOTION INSTEAD OF POSTURE

For too long designers have employed the simplistic idea that humans will sit or stand for long periods in the same body posture with little or no variation — for example, while keyboarding in the office, driving a car, operating a machine, or working at a check-out counter. This illusionary concept implies that in each case, there is only one good, healthy, or correct posture that we all always assume, or should assume, while sitting or standing. To some extent this false concept of the one and only "right" body stance may have been generated by the standardized erect sitting or standing postures employed for measuring body sizes (see the section Design to Fit Body Dimensions, below).

Probably the upright posture has been the prevalent drawing guide, mostly because the designer can easily visualize it and use it as a design template. But how did this concept originate? Thousands of years ago, Egyptian drawings showed mostly stiff, upright sitting postures of goddesses, gods, and rulers, but we do not know what their everyday sitting postures actually were. Workers, however, were depicted in motion. Grave steles in the National Archeological Museum in Athens indicate that about 2,500 years ago, ancient Greeks used several kinds of seats. Some were upright with headrests at the top of straight vertical backs. Other steles depict curved chairs with reclining backs equipped with contoured back supports. In some cases, footrests helped support the seated person comfortably.

The orthopedist Staffel and his colleagues introduced the erect (upright) posture in the 1880s as "healthy" — but he meant this label to indicate a contrast with the sickly, bent, and hunched poses of the generally malnourished tailors and farm workers of that time (Staffel, 1884, 1889). This led to "The erect posture is the healthy posture" motto that for about a hundred years misguided the design of furniture, especially for schoolchildren and office workers (Bonne, 1969; K. H. E. Kroemer & A. D. Kroemer, 2001a; Tenner, 2003b; Zacharkow, 1988). A newer buzzword is the ill-defined "neutral" posture. Simply eliminating the word *posture* in the design context and instead using *motion* would help to overcome outdated and misleading design customs (Lueder & Noro, 1994).

Humans are unable to maintain any given posture over long periods. When standing still, sitting immobile, or lying quietly, we quickly feel uncomfortable, and over time it becomes almost impossible to sustain the posture. Even when sleeping, we do not lie stiff and still but frequently move around. If injury or disease imposes a fixed body position, circulatory and metabolic functions become impaired; people who must lie in place develop bedsores. Apparently, the human body is made to move, not to stay still. This means that we must design for change and for motion. Publications by Lueder and Noro (1994), K. H. E. Kroemer and A. D. Kroemer, (2001a, 2001b), K. H. E. Kroemer, H. B. Kroemer, and Kroemer-Elbert (2001), and the draft U.S. standard on computer workstations (HFES, 2002) reflect this new ergonomics thinking about the design of the sit-down workplace for adults, adolescents, or children.

Our bodies are designed for movement especially in the legs and arms. Strong legs are able to propel the body on the ground, with most motion occurring in the knee and hip joints. Our thumbs and fingers are able to perform finely controlled and complex motions, and the elbow and shoulder joints provide extensive angular freedom. Of course, overdoing motions can lead to trouble. Since the early 1700s, wrist problems have been associated with excessive movement requirements (Armstrong, 2000; Kroemer, 2001; Violante, Isolani, & Raffi, 2000). The head and neck can bend and twist. Movements of the trunk occur mostly in flexion and extension at the lower back. However, these bending motions are limited and often lead to overexertion if combined with sideways rotation of the torso; low back pain has been reported throughout the history of humankind (Snook, 2001).

Designing to fit suitable motion ranges instead of fixed postures is not difficult. Known degrees of freedom of movement apply to the articulations in the human

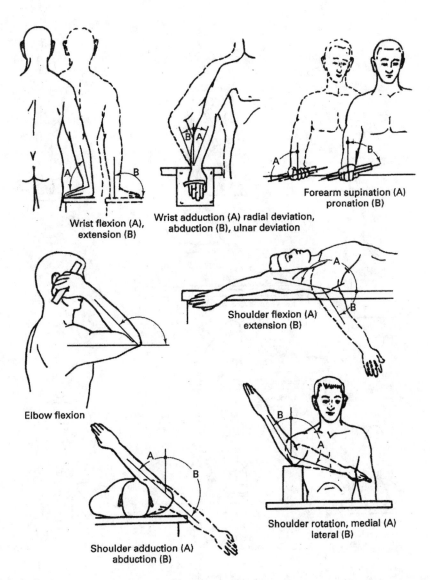

Wrist flexion (A),
extension (B)

Wrist adduction (A) radial deviation,
abduction (B), ulnar deviation

Forearm supination (A)
pronation (B)

Elbow flexion

Shoulder flexion (A)
extension (B)

Shoulder adduction (A)
abduction (B)

Shoulder rotation, medial (A)
lateral (B)

FIGURE 4.1 Displacements in body joints (continued on page 82). (Adapted from *Human Engineering Guide to Equipment Design,* by H. P. Van Cott and R. G. Kinkade, 1972, Washington, DC: U.S. Government Printing Office.)

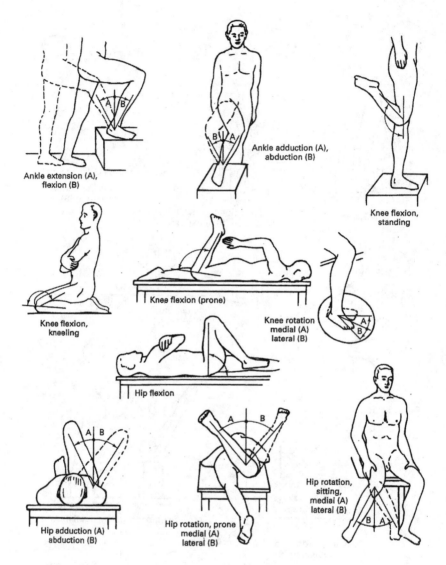

FIGURE 4.1 (continued)   Displacement in body joints.

body, as shown schematically in Figure 4.1 for major body joints. Of course, ranges of motion (also called *mobility* or *flexibility*) depend greatly on age, health, fitness, training, skill, and any disability.

In the past, studies of mobility employed dissimilar measuring instructions and techniques to assess flexibility ranges of various groups of people (Wu et al., 2002); hence, there is much diversity in reported results. However, at least one set of mobility measurements has been taken on groups of 100 females and 100 males by the same researchers using the same techniques. These data are reported in Table 4.1. Note that for design purposes, the small existing differences in mobility between the groups of males and females are negligible.

Students of physical education in the groups studied demonstrated the maximal mobility ranges shown in Table 4.1; most ordinary people will have slightly less flexibility. Preferred ranges may differ according to habits, skills, and strength requirements. "Convenient" mobility is somewhere within the array of maximal values listed in the table but not always in the middle of the range. Many persons with disabilities are not at all mobile; special assessments, described in the next chapter, ascertain their specific motion capabilities and deficiencies.

Design for motion starts by establishing the actual movement ranges. Convenient motions may cluster around the mean mobility of a body joint, but often they are close to its limits. For example, in a person walking around on a job site or standing, most of the time the knees are almost fully extended. That is, in the sagittal view, the actual knee angles are almost at the extreme value of about 180°. In the view from the side, the hip angle (between trunk and thigh) also varies in the neighborhood of 180°. However, when sitting, both angles cluster around 90°.

Table 4.2 lists motion ranges associated with everyday activities. Given the many kinds of motion deficiencies, similar tables for "extra-ordinary" persons as groups are missing in the general literature but could be compiled when sufficient data become available.

Preferred work areas of the hands and feet are in front of the body, within curved envelopes that reflect the mobility of the arms and legs. The envelope for manipulation is primarily determined by movement of the forearm in the elbow joint and of the whole arm in the shoulder joint. The movement area for the foot depends on motions of the lower leg in the knee joint and of the total leg in the hip joint. Thus, these reach envelopes for hands and feet are often described as partial spheres around the presumed locations of the body joints. However, the preferred ranges within the possible motion zones are different when the main requirements are strength, speed, accuracy, or vision, as discussed in some detail by Kroemer (1997a,b), and Kroemer et al. (2001). Clearly, there is not just one reach envelope; rather, different people prefer different envelopes for different tasks.

For each job situation, the ergonomist determines the dominant requirements of the task; for example, when the operator

- works while sitting, walking, or standing
- performs wide-ranging or specialized work
- must exert large or small forces
- executes fast and gross or slow and exact motions
- needs high or low visual control.

All these circumstances affect the selection of the specific work envelope.

TABLE 4.1
Comparison of Mobility Data (in Degrees) for Females and Males

Joint	Movement	5th Percentile		50th Percentile		95th Percentile		Difference Female Minus Male Values
		Female	Male	Female	Male	Female	Male	
Neck	Ventral flexion	34.0	25.0	51.5	43.0	69.0	60.0	+8.5
	Dorsal flexion	47.5	38.0	70.5	56.5	93.5	74.0	+14.0
	Right rotation	67.0	56.0	81.0	74.0	95.0	85.0	+7.0
	Left rotation	64.0	67.5	77.0	77.0	90.0	85.0	None
Shoulder	Flexion	169.5	161.0	184.5	178.0	199.5	193.5	+6.5
	Extension	47.0	41.5	66.0	57.5	85.0	76.0	+8.5
	Adduction	37.5	36.0	52.5	50.5	67.5	63.0	NS
	Abduction	106.0	106.0	122.5	123.5	139.0	140.0	NS
	Medial rotation	94.0	68.5	110.5	95.0	127.0	114.0	+15.5
	Lateral rotation	19.5	16.0	37.0	31.5	54.5	46.0	+5.5
Elbow	Flexion	135.5	122.51	148.0	138.0	160.5	150.0	+10.0
	Supination	87.0	86.0	108.5	107.5	130.0	135.0	NS
	Pronation	63.0	42.5	81.0	65.0	99.0	86.5	+16.0
Wrist	Extension	56.5	47.0	72.0	62.0	87.5	76.0	+10.0
	Flexion	53.5	50.5	71.5	67.5	89.5	85.0	+4.0
	Adduction	16.5	14.0	26.5	22.0	36.5	30.0	+4.5
	Abduction	19.0	22.0	28.0	30.5	37.0	40.0	2.5
Hip	Flexion	103.0	95.0	125.0	109.5	147.0	130.0	+15.5
	Adduction	27.0	15.5	38.5	26.0	50.0	39.0	+12.5
	Abduction	47.0	38.0	66.0	59.0	85.0	81.0	+7.0
	Medial rotation (prone)	30.5	30.5	44.5	46.0	58.5	62.5	NS
	Lateral rotation (prone)	29.0	21.5	45.5	33.0	62.0	46.0	+12.5
	Medial rotation (sitting)	20.5	18.0	32.0	28.0	43.5	43.0	+4.0
	Lateral rotation (sitting)	20.5	18.0	33.0	26.5	45.5	37.0	+6.5
Knee	Flexion (standing)	99.5	87.0	113.5	103.5	127.5	122.0	+10.0
	Flexion (prone)	116.0	99.5	130.0	117.0	144.0	130.0	+13.0
	Medial rotation	18.5	14.5	31.5	23.0	44.5	35.0	+8.5
	Lateral rotation	28.5	21.0	43.5	33.5	58.5	48.0	+10.0
Ankle	Flexion	13.0	18.0	23.0	29.0	33.0	34.0	−6.0
	Extension	30.5	21.0	41.0	35.5	51.5	51.5	+5.5
	Adduction	13.0	15.0	23.5	25.0	34.0	38.0	NS
	Abduction	11.5	11.0	24.0	19.0	36.5	30.0	+5.0

Adapted from "A Comparison of Range of Joint Mobility in College Females and Males," by K. R. Staff, 1983, unpublished master's thesis, Texas A&M University, College Station, TX.

Note: The Difference column lists only those at the 50th percentile and if significant ($\alpha < 0.5$); NS = not significant.

TABLE 4.2
Estimates of Actual Motion Ranges at Work

Angles	Walking About, Standing	Sitting
Knee (in the lateral view)	Moving between about 150° and full stretch, which is at approximately 180°	Moving about ± 30° about midrange, which is at approximately 90°
Hip (in the lateral view)	Moving between about 170° and full stretch, which is at approximately 180°	Moving about ± 15° about midrange, which is at approximately 90°
Shoulder (in all views)	Moving about ± 30° about midrange, which is the upper arm hanging down	
Elbow (in the lateral view)	Moving about ± 30° about midrange, which is at approximately 90°	
Wrist (in all views)	Moving about ± 15° about midrange, which is in line with the forearm	
Neck/head (in all views)	Moving about ± 10° about midrange, which is in line with the trunk	
Back (in all views)	Moving about ± 10° about midrange, which is erect	

Adapted from *Engineering Physiology: Bases of Human Factors/Ergonomics* (3rd ed.), by K.H.E Kroemer, H.J. Kroemer, and K.E. Kroemer-Elbert, 1997, New York: Van Nostrand Reinhold–Wiley. Reprinted with permission of John Wiley & Sons, New York. All rights reserved.

Designing for movement follows these steps:

Step 1: Determine the major body segments and joints involved.

Step 2: Adjust body dimensions reported for standardized postures (such as in the tables in the Appendix) *to accommodate the real conditions.* Table 4.3 provides guidance.

Step 3: Select appropriate motion ranges in the body joints. The range can be depicted as the area between two positions, such as knee angles between 60 and 90°; or as a motion envelope, such as circumscribed by combined hand-and-arm movements; or a clearance envelope under or within which body parts must fit; or an envelope through and beyond which body parts must reach.

Step 4: Design and test; modify as necessary.

Avoid the following postures:

Avoid all postures that must be maintained for long periods, especially at the extreme limits of the range of motion. This is particularly important for the wrist and back.

Avoid twisted body positions, especially of the trunk and neck. Twists result often from bad location of work objects, controls, and displays, especially when placed off to the side.

Avoid forward bending of the trunk, neck, and head. Improperly positioned controls and visual targets, including working surfaces that are too low, frequently cause such bending.

Avoid holding the arms raised. An elevated-arm posture results commonly from locating controls or objects too high. Manipulation is convenient at elbow height or slightly above when the upper arm hangs down.

TABLE 4.3
Guidelines for Workspace Design

Consider this item —	In order to
• Human strength	Facilitate exertion of strength (work, power) by object location and orientation
• Human speed	Place items so that they can be reached and manipulated quickly
• Human effort	Arrange work so that it can be performed with least effort
• Human accuracy	Select and position objects so that they can be manipulated and seen with ease
• Importance	Place the most important items in the most accessible locations
• Frequency of use	Place the most frequently used items in the most accessible locations
• Function	Group items with similar functions together
• Sequence of use	Lay out items which are commonly used in sequence in that sequence

Adapted from *Engineering Physiology: Bases of Human Factors/Ergonomics* (3rd ed.), by K. H. E. Kroemer, H. J. Kroemer, and K. E. Kroemer-Elbert, 1997, New York: Van Nostrand Reinhold–Wiley. Reprinted with permission of John Wiley & Sons, New York. All rights reserved.

Table 4.3 provides guidelines for workspace design.

DESIGN TO FIT BODY DIMENSIONS

To assess a person's body size, or strength, it is generally best to follow standard anthropometric procedures so that the individual's makeup can be compared with population data; for example, with the anthropometric data listed in the Appendix to this book or the strength data collected by Kumar (2004). Even if a person's traits or tasks diverge so much from the usual that specific measurements are called for, it is usually advantageous to employ the classic measurements as far as is feasible, just to allow comparisons. Note, however, that new 3-D techniques (Landau, 2000; Robinette, 2000) may change measurement processes, data storage, and use of anthropometric data in the near future (see Chapters 2 and 7).

For standard anthropometry, the body is put into one of the two basic static measuring positions:

Standing: The classic instruction is to stand erect, heels together (sometimes with the rear of heels, buttock, and shoulders touching a vertical wall), head erect, look straight ahead, arms hanging straight down (or upper arms hanging, forearms horizontal and extended forward), fingers extended.

Sitting: The classic instruction is (while the person sits on a plain, horizontal, hard surface adjusted in height so that the thighs are horizontal, with lower legs vertical, feet flat on the floor) to hold trunk and head erect, look straight

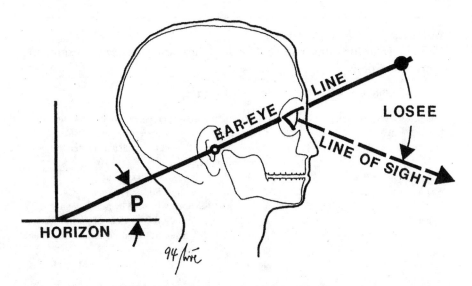

FIGURE 4.2 The ear–eye line passes through the right ear hole and the juncture of the lids of the right eye. The EE line serves as the reference for head posture: the head is held upright (erect) when the tilt angle *P* is about 15° above the horizon. The EE line is also the reference for the line-of-sight angle, LOSEE. (From *Engineering Physiology: Bases of Human Factors/Ergonomics* (3rd ed.), by K. H. E. Kroemer, H. B. Kroemer, and K. E. Kroemer-Elbert, 1997, New York: Van Nostrand Reinhold/Wiley. Reprinted with permission of John Wiley & Sons, New York. All rights reserved.)

ahead, arms hanging straight down (or upper arms hanging, forearms horizontal and extended forward), fingers extended.

Whether the person stands or sits, the head (including the neck) is held erect (upright) when, in the front view, the pupils are aligned horizontally, and, in the side view, the ear-eye line is tilted about 15° above the horizon (see Figure 4.2).

Under everyday conditions at work or home, people do not stand or sit in these standard positions naturally. Thus, we must convert the dimensions taken on the body in the measuring poses to reflect real postures. The postures assumed at work, at home, or at leisure can and do vary greatly. Therefore, no simple conversion factors exist that apply to all conditions. For each application, the designer must estimate the corrections that apply to the anticipated postures. Table 4.4 contains general guidelines for doing this task.

Of course, we really intend to design for body movements. The outputs of motion capture systems, currently used mostly in rehabilitation laboratories, are often somewhat cumbersome to apply for design purposes (Seitz et al., 2003). The emerging 3-D anthropometry (Robinette, 2000), briefly described in the preceding chapters, allows — at least in theory — measurement of the human body in motion. When this new technique becomes widely used, it should bring about a new approach to dynamic design.

TABLE 4.4
Guidelines for the Conversion of Standard Measuring Postures to Common Work Conditions

To consider this item —	Do the following
Slumped standing or sitting	Deduct 5 to 10% from appropriate height measurements.
Relaxed trunk	Add 5 to 10% to trunk circumferences and depths.
Wearing shoes	Add approximately 25 mm to standing and sitting heights; more for high heels.
Wearing light clothing	Add about 5% to appropriate dimensions.
Wearing heavy clothing	Add 15% or more to appropriate dimensions. (Note that mobility may be strongly reduced by heavy clothing.)
Extended reaches	Add 10% or more for extensive motions of the trunk.
Use of hand tools	Center of handle is at about 40% hand length, measured from the wrist.
Comfortable seat height	Add or subtract up to 10% to or from standard seat height.

Adapted from *Engineering Physiology: Bases of Human Factors/Ergonomics* (3rd ed.), by K. H. E. Kroemer, H. J. Kroemer, and K. E. Kroemer-Elbert, 1997, New York: Van Nostrand Reinhold–Wiley. Reprinted with permission of John Wiley & Sons, New York. All rights reserved.

The body dimensions that are commonly taken in classic anthropometric surveys are illustrated in Figures 4.3 to 4.10 and described in the following text. (The numbers in brackets are those employed by Gordon et al. (1989), who also provide precise definitions of the anthropometric terms. The numbering is the same as used in Tables A.2 through A.10 in the Appendix.)

1. STATURE [99]

The vertical distance from the floor to the top of the head when standing. The main reference for comparing population samples; it relates to the minimal height for clearance of overhead obstructions. For design, add height for more clearance for hat, shoes, and stride; conversely, consider slump when standing and in motion.

2. EYE HEIGHT, STANDING [D19]

The vertical distance from the floor to the outer corner of the right eye when standing. Origin of the visual field; reference point for the location of visual targets such as displays and for vision obstructions. For design, consider slump and motion of the standing person.

3. SHOULDER HEIGHT (ACROMION), STANDING [2]

The vertical distance from the floor to the tip of the shoulder (acromion), when standing. The acromion is near the shoulder joint, the center of rotation of the upper arm; starting point for arm length measurements, reference point for hand reaches. For design, consider slump and motion of the standing person.

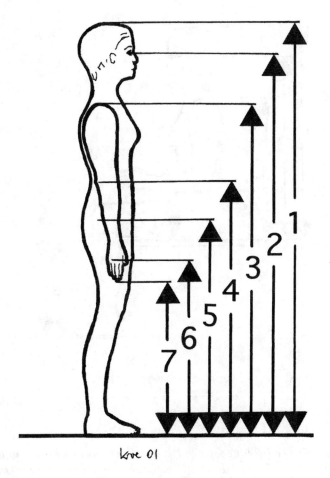

FIGURE 4.3 Heights measured on a person standing upright.

4. ELBOW HEIGHT, STANDING [D16]

The vertical distance, when standing, from the floor to the lowest point of the right elbow, which is flexed at 90°. Reference point for height and distance of the work area of the hand and for location of controls and fixtures. For design, consider slump and motion of the standing person.

5. HIP HEIGHT (TROCHANTER), STANDING [107]

The vertical distance from the floor to the trochanter landmark on the upper side of the right thigh, when standing. Starting point, near the center of the hip joint, for many leg length measurements; reference point for leg reaches. For design, consider slump and motion of the standing person.

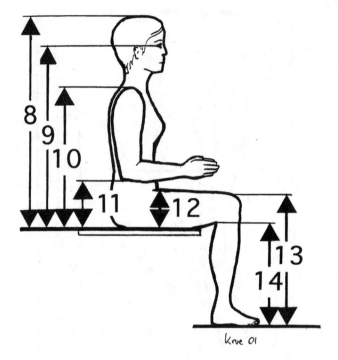

FIGURE 4.4 Heights measured on a person sitting upright.

6. KNUCKLE HEIGHT, STANDING

The vertical distance from the floor to the knuckle (metacarpal bone) of the middle finger of the right hand, when standing. Reference point for low locations of controls, handles, and handrails. For design, consider slump and motion of the standing person.

7. FINGERTIP HEIGHT, STANDING [D13]

The vertical distance from the floor to the tip of the middle finger of the right hand, when standing. Reference point for the lowest location of controls, handles, and handrails. For design, consider slump and motion of the standing person.

8. SITTING HEIGHT [93]

The vertical distance from the seat surface to the top of the head, when sitting. Relates to the minimal height of overhead obstructions. For design, add height for more clearance for hat and trunk motion; conversely, consider slump and motion of the sitting person.

9. SITTING EYE HEIGHT [49]

The vertical distance from the seat surface to the outer corner of the right eye, when sitting. Origin of the visual field and reference point for the location of visual targets

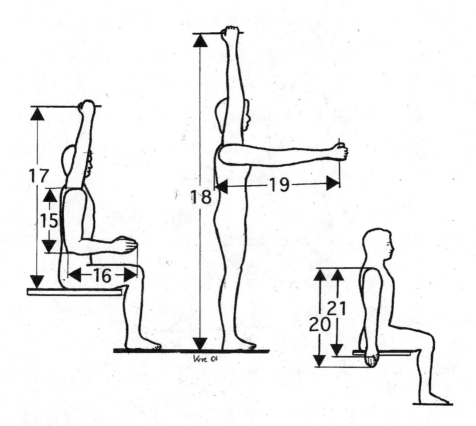

FIGURE 4.5 Arm and reach measurements.

such as displays and vision obstructions. For design, consider slump and motion of the seated person.

10. Sitting Shoulder Height (Acromion) [3]

The vertical distance from the seat surface to the tip of the shoulder (acromion), when sitting. The acromion is near the shoulder joint, the center of rotation of the upper arm; starting point for arm length measurements and reference point for hand reaches. For design, consider slump and motion of the seated person.

11. Sitting Elbow Height [48]

The vertical distance from the seat surface to the lowest point of the right elbow, when sitting, with the elbow flexed at 90°. Reference point for height of an armrest, of the work area when use of the hand is involved, and of keyboard and controls. For design, consider slump and motion of the seated person.

12. Sitting Thigh Height (Clearance) [104]

The vertical distance from the seat surface to the highest point on top of the right thigh, when sitting, with the knees flexed at 90°. Minimal clearance needed between

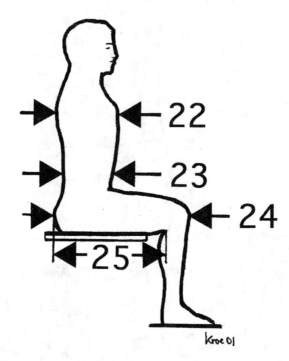

FIGURE 4.6 Measurements of body depths.

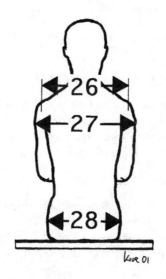

FIGURE 4.7 Measurements of body widths.

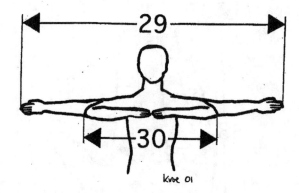

FIGURE 4.8 Measurements of arm spans.

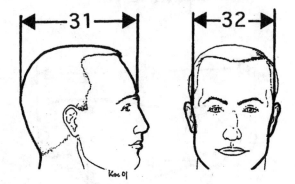

FIGURE 4.9 Head measurements.

seat pan and the underside of a structure, such as a table. For design, add clearance for clothing and motions.

13. Sitting Knee Height [73]

The vertical distance from the floor to the top of the right kneecap, when sitting, with the knees flexed at 90°. Minimal clearance needed below the underside of a structure, such as a table. For design, add height for shoe and motions.

14. Sitting Popliteal Height [86]

The vertical distance from the floor to the underside of the thigh directly behind the right knee, when sitting, with the knees flexed at 90°. Reference for the height of a seat. For design, add height for shoes, or reduce height to avoid pressure at the front edge of the seat pan; consider movement of the feet.

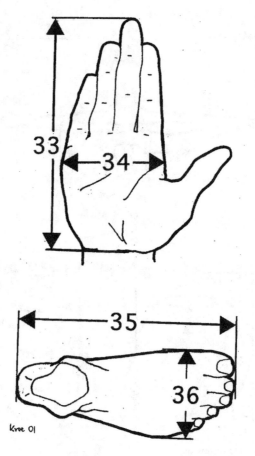

FIGURE 4.10 Hand and foot measurements.

15. Shoulder–Elbow Length [91]

The vertical distance from the underside of the right elbow to the right acromion, with the elbow flexed at 90° and the upper arm hanging vertically. A general reference for comparing population samples.

16. Elbow–Fingertip Length [54]

The horizontal distance from the back of the right elbow to the tip of the middle finger, with the elbow flexed at 90° and wrist and fingers straight. Reference for fingertip reach when moving the forearm around the elbow joint. For design, add length for motions but reduce length if an object is to be held in the palm.

17. Overhead Grip Reach, Sitting [D45]

The vertical distance from the seat surface to the center of a cylindrical rod firmly held in the palm of the right hand, with the right arm and wrist extended upward.

Reference for height of overhead controls to be operated by a seated person. For design, consider ease of motion, kind of reach, and finger/hand/arm strength.

18. OVERHEAD GRIP REACH, STANDING [D42]

The vertical distance from the standing surface to the center of a cylindrical rod firmly held in the palm of the right hand, with the right arm and wrist extended upward. Reference for height of overhead controls to be operated by a standing person. Add shoe height. For design, consider ease of motion, kind of reach, and finger/hand/arm strength.

19. FORWARD GRIP REACH [D21]

The horizontal distance from the back of the right shoulder blade to the center of a cylindrical rod firmly held in the palm of the right hand, with the right arm and wrist extended forward. Reference for forward reach distance. For design, consider ease of motion, kind of reach, and finger/hand/arm strength.

20. ARM LENGTH, VERTICAL [D3]

The vertical distance from the tip of the right middle finger to the right acromion, with the arm hanging vertically. A general reference for comparing population samples. Reference for the location of controls very low on the side of the operator. For design, consider ease of motion, kind of reach, and finger/hand/arm strength.

21. DOWNWARD GRIP REACH [D43]

The vertical distance from the right acromion to the center of a cylindrical rod firmly held in the palm of the right hand, with the arm hanging vertically. Reference for the location of controls low on the side of the operator. For design, consider ease of motion, kind of reach, and finger/hand/arm strength.

22. CHEST DEPTH [36]

The horizontal distance from the back to the right nipple. A general reference for comparing population samples; reference for the clearance between seat backrest and the location of obstructions in front of the trunk. For design, add space for ease of motion.

23. ABDOMINAL DEPTH, SITTING [1]

The horizontal distance from the back to the most protruding point on the abdomen. A general reference for comparing population samples; reference for the clearance between seat backrest and the location of obstructions in front of the trunk. For design, add space for ease of motion.

24. BUTTOCK-KNEE DEPTH, SITTING [26]

The horizontal distance from the back of the buttocks to the most protruding point on the right knee, when sitting with the knees flexed at 90°. Reference for the clearance between seat backrest and the location of obstructions in front of the knees. For design, add space for ease of motion.

25. BUTTOCK-POPLITEAL DEPTH, SITTING [27]

The horizontal distance from the back of the buttocks to the back of the right knee just below the thigh when sitting with the knees flexed at 90°. Reference for the depth of a seat. For design, add space for ease of motion.

26. SHOULDER BREADTH, BIACROMIAL [10]

The horizontal distance between the right and left acromia. A general reference for comparing population samples. Somewhat longer than, but indicative of the distance between the centers of rotation of the upper arms (shoulder joints). For design, add space for ease of motion.

27. SHOULDER BREADTH, BIDELTOID [12]

The maximal horizontal breadth across the shoulders between the lateral margins of the right and left deltoid muscles. Reference for clearance requirement at shoulder level. For design, add space for ease of motion.

28. HIP BREADTH, SITTING [66]

The maximal horizontal breadth across the hips or thighs, whichever is greater, when sitting. Reference for seat width. For design, add space for clothing and ease of motion.

29. SPAN [98]

The horizontal distance between the tips of the middle fingers of the horizontally outstretched arms and hands. Reference for reach to the sides. For design, consider ease of reach and motion.

30. ELBOW SPAN (ARMS AKIMBO)

The horizontal distance between the tips of the elbows of the horizontally outstretched upper arms with the elbows flexed so that the fingertips of both hands meet in front of the trunk. Reference for "elbow room." For design, consider motion.

31. HEAD LENGTH [62]

The horizontal distance from the glabella (the center of the browridges) to the most rearward protrusion in the middle of the back of the skull (the occiput). A general reference for comparing population samples; reference for headgear size.

32. HEAD BREADTH [60]

The maximal horizontal breadth of the head above the attachment of the ears. A general reference for comparing population samples; reference for headgear size.

33. HAND LENGTH [59]

The length of the right hand between the crease of the wrist and the tip of the middle finger, with the hand flat. A general reference for comparing population samples; reference for size of hand tools and gears. For design, consider changes due to manipulations and use of gloves and tools.

34. HAND BREADTH [57]

The breadth of the right hand across the knuckles of the four fingers. A general reference for comparing population samples; reference for size of hand tools and gears and the opening through which a hand may fit. For design, consider changes due to manipulations and use of gloves and tools.

35. FOOT LENGTH [51]

The maximal horizontal length of the right foot when standing. A general reference for comparing population samples; reference for shoe and pedal size.

36. FOOT BREADTH [50]

The maximal horizontal breadth of the right foot, at a right angle to the long axis of the foot, when standing. A general reference for comparing population samples; reference for shoe size and spacing of pedals.

37. WEIGHT

Weight of the nude body taken to the nearest tenth of a kilogram. A general reference for comparing population samples; reference for body size, clothing, strength, health, and so forth. For design, add weight for clothing and equipment worn on the body.

DESIGN FOR VERY SMALL AND VERY BIG PEOPLE

People come in greatly differing body sizes, as is apparent if we compare people from around the world (see Table A.1 in the Appendix). However, wide variety exists even within any one of the relatively homogeneous population groups — say, Italians or Texans. The immense variability applies not only to overall size but also to the way in which body segments are combined. There are tall people with thin bodies and big feet; there are short persons who are heavy and strong. Two individuals can have the same stature, but whereas one may have long legs and a short trunk, the other could be short-legged and long-bodied.

WHAT DATA TO USE?

Everyday experience confirms what anthropometrists and statisticians have told us for decades: There are no "single-percentile" humans, either of the average (which means 50th percentile) or of any other single value, including the often invoked mysterious 5th and 95th percentile phantoms (see Chapter 3). Therefore, to repeat what has been said before, anything designed for the illusory "Mrs. and Mr. Average" in reality fits nobody well; neither do designs for the even more illusory "Miss Fifth-Percentile" and "Mr. Ninety-Five" — all such generalities being figments of the imagination.

The designer strives to fit ranges of body sizes. For this goal, first those body dimensions that are critical to the design must be determined. Examples are inter-pupillary distance for eyeglasses, head width and length for helmets, reach lengths and trunk angles for work spaces, and leg angles and foot sizes for pedals. Then numerical values must be obtained for those dimensions that determine the accom-modation of tasks and objects (Moroney & Cameron, 2001). The HFES 300 Com-mittee (2004) provided detailed guidelines for using anthropometric data in design.

CALCULATING DATA

With luck, values for design data that we look for would appear in an existing table of anthropometric data, such as listed in the Appendix. Otherwise, we shall have to calculate the needed numbers. We work out the percentile values from the mean and the standard deviation by applying k factors selected from Table 3.1; Chapter 3 contains a detailed description of how to do this. (These calculations are valid only for Gaussian distributions.)

However, we must apply great caution when calculating a value that is at the extreme end of a data distribution, the reason being that the "smoothing" of the data (which occurs automatically as part of statistical procedures that presume a Gaussian distribution) is likely to hide outliers in the actual data. Similarly, we must be careful when using correlation coefficients (as compiled in Table 3.2) to calculate extreme values (that is, at the tails of distributions). We need to keep in mind that the commonly used correlation coefficient is a statistical number calculated under the (usually unchecked and possibly false) assumption of a straight-line regression between two sets of data. The assumption of a straight-line relation applies over the whole length of the distributions. Yet, raw data are often scattered relatively more at the tails than in the middle of a distribution. Thus, using the common statistical procedures (described in Chapter 3) to calculate the values of outliers at the tails of distributions is seldom a reliable approach (Gordon, Corner, & Brantley, 1997).

Trying to design for "extra-ordinary" people, we realize that most are, indeed, outliers from the overall population statistics. Therefore, for practical and statistical reasons, we must generally measure special persons' dimensions individually. More about this follows.

An example can demonstrate how to proceed and why care is needed. Check the body weight of U.S. soldiers in Table A.10 of the appendix. For females, it lists an average of 62 kg with a standard deviation of 8 kg. The corresponding values for males are 79 kg and 13 kg. Calculating the 95th percentile (with $k = 1.64$, from Table 3.2) yields 75.1 kg (a bit more than 165 lb) for females and 100.3 kg (about 221 lb) for males. A casual observation of American customers at a grocery store, spectators at a ball game, train or bus passengers at a station, or people at just about any other locale reveals that more than merely 5% of the people present appear as heavy or even heavier. Unfortunately, recent statistics about the rapidly increasing numbers of overweight and even obese adults and children confirm this observation (Cole, Belizzi, Flegal, & Dietz, 2000; Ogden et al., 2004; Strauss & Pollack, 2001).

One (correct) argument is that soldiers are not representative of the general civilian population, but we have no choice but to use the military data because reliable information on civilians is not available. If we tried to use ratio scaling or regressions (see Chapter 3), we would run into the problem of correlations: In the Army sample, correlation between stature and weight was about 0.5 for both genders (Cheverud et al., 1990). In the general population, the correlation would be much lower, around 0.3, possibly even less.

Fontaine, Gadbury, Heymsfield, Kral, Albu, and Allison (2002) used linear regression analysis to calculate how body dimensions of morbidly obese people depend on body weight. However, given the uncertainty about the tails of anthropometric distributions in general and the frequent need to assess the traits of a few special individuals, one often has to resort to actually measuring the body dimensions of interest, rather than calculating them from existing statistics (see M. H. Al-Haboubi, 1999).

MEASURING DATA

The process of first establishing groups of people who possess unusual body dimensions, then setting up proper sampling procedures, and finally measuring samples is very costly and time consuming for some obvious and several hidden reasons (Kalton & Anderson, 1989). If that is the chosen procedure, one places the body in standardized positions according to the rules of traditional anthropometry and takes the measurements of interest, as described earlier. Guidance is provided in texts by Cheverud et al. (1990); Garrett and Kennedy (1971); Gordon et al. (1989); Landau (2000); Lohman, Roche, and Martorel (1988); NASA/Webb (1978); and Roebuck (1995). Keep in mind, however, that measurements taken on persons in the standard erect postures (standing or sitting) must be adjusted to reflect the actual poses of the body, as discussed earlier in this chapter. Furthermore, certain groups of "extra-ordinary" people may not be able to assume upright postures.

A good reason for choosing to do the assessment in standard ways is to obtain information that is easily comparable with the general data sets. In other cases, we measure these individuals in nonstandard ways and possibly use new measurement techniques (as recently discussed, for example, by Landau, 2000; Feathers, Polzin, Paquet, Lenker, & Steinfeld, 2001; M. J. J. Wang, E. M. Y. Wang, & Lin, 2002a, 2002b) to obtain the exact information we need to fit their special needs. An example is the determination of reach capabilities of a person who, because of a back injury, cannot stand upright.

We may hope that emerging computerized 3-D techniques will finally facilitate measurement of the moving body and processing the measurements into computerized design models that prove to be anatomically, biodynamically, and ergonomically correct and easy to use. For the time being, however, we must measure individual weights and hip breadths to design chairs and seats in automobiles, buses, and aircraft that accommodate large people, for instance. Likewise, we must individually measure the eye heights of very short and very tall people, standing and sitting, to suitably locate the display on a ticket vending machine or an ATM and on the monitor of a desktop computer.

DESIGN TO AVOID HARM AND INJURY

Most people may not appreciate the economic implications and associated design efforts, but entire industries depend on making easy chairs and wheelchairs, producing beds and gurneys, constructing and maintaining hospitals and homes for the invalid and old, and designing reliable and safe instruments to diagnose and monitor body functions and gadgets to dispense medications. Physicians, psychologists, rehabilitation specialists, therapists, nurses and their assistants and aides, and other professionals care for and deal with ill, injured, and bedridden patients. We hope not to become a patient by illness or injury, and yet misfortune can happen at any time, and we may have to stay in a wheelchair or lie in bed for a while, perhaps a long time.

MEDICAL ERRORS

As early as 1970, Rappaport succinctly affirmed an urgent need for better ergonomic solutions in health care. Among the first human engineering guidelines were those developed in 1988 by the Association for the Advancement of Medical Instrumentation (AAMI). Succinct recommendations were derived, largely from already well established U.S. military standards, especially MIL-STD 1472. In 1992, however, the U.S. government documented nearly 8,000 cases that occurred since 1984 in which patients had been injured or killed due to human error in the use of therapeutic and medical devices (Wiklund & Smith, 1992). Even a decade later, about the same number of deaths and injuries happened due to medical error, a sorry sign that the old problem still had not been solved (Woods, 2000).

Among the primary ergonomics issues are errors in the use of diagnostic and therapeutic equipment. Improper operation of medical devices is frequent and so is the misinterpretation of diagnostic information from complex instruments such as imaging devices. These topics are of great interest to all involved: patients, physicians and other caregivers, administrators, and attorneys. Involving ergonomists continuously in the design process, from the initial concept through construction and testing, should yield better design of all technical instruments from simple devices to computerized expert systems. This is particularly important with equipment meant for rapid, mobile emergency deployment, such as in first aid and treatment of accident victims, when both the patients and the caregivers often are under great stress.

Many of the related human engineering topics are beyond the scope of this text, and the reader wanting more information would best contact appropriate professional, industrial, and trade organizations. In the United States, the Human Factors and Ergonomics Society has technical committees concerned with several of these issues. Furthermore, there is substantial literature available, for example by Rice (1998), Megaw (2001), and W. A. Rogers, Mykittyshin, Campbell, and Fisk (2001).

Repetition-Caused Injuries

Repetitive injuries to muscle, other connective tissues, and the joints of the human body have become a major problem in occupational safety and health. In 1713, the Italian physician Bernadino Ramazzini described occupation-related diseases and injuries and musculoskeletal disorders in professions and trades. (In 1993, Wright provided an easily readable translation of the original Latin text; and in 2004, Franco and Fusetti reviewed Ramazzini's observations.) Ramazzini associated most occupational illnesses and accidents with physical labor but described how even secretaries and scribes were known to suffer from muscle pain and cramps, which Ramazzini explained as stemming from repetitive tasks that involved long-maintained hand and arm postures and often-repeated motions.

In the mid-1800s, certain occupations and repetitive strain disorders were said to be correlated; examples are textile workers (Burry & Stoke, 1985) and musicians, especially pianists (Hochberg, Leffert, Heller, & Merriman, 1983; Fry, 1986a, 1986b, 1986c, 1986d, 1986e; Lockwood, 1989). The pianists' health problems became a concern to keyboard users in offices around 1900. In the 1860s–1870s, Sholes and his coinventors had obtained several U.S. patents for mechanical "Type-Writers." At first, the typewriter keyboards resembled the key arrangements of the piano. However, Shole's last and ultimately successful patent (No. 207,559 of 1878; see Sholes, 1878) showed four straight rows of 11 round key buttons. The third row, counted from the operator, has the letters QWERTY inscribed on the left-side keys, an arrangement still used on today's computer keyboards in English. Obviously, and unfortunately, Shole's QWERTY keyboard was not ergonomically designed, which is probably one of the reasons for repetitive injuries to keyboard users. (For more information, see Kroemer, 2001; K. H. E. Kroemer & A. D. Kroemer, 2001a).

By the 1890s, the literature had begun to describe a variety of occupation-related repetitive diseases. The factors that cause, aggravate, or precipitate connective tissue disorders (cumulative tissue disorders, CTDs) can be part of occupational or leisure activities, as indicated by the names of disorders such as

- writer's cramp or scribe's palsy
- telegraphist's or pianist's (or stitcher's, tobacco-primer's, meat cutter's, washerwoman's) wrist
- goalkeeper's, seamstress,' or tailor's finger
- bowler's or gamekeeper's or jeweler's thumb
- bricklayer's hand
- jackhammerer's or carpenter's arm
- carpenter's or jailer's or student's elbow
- porter's neck
- shoveller's hip
- weaver's bottom
- housemaid's or nun's knee
- ballet dancer's or nurse's foot.

In the 20th century, new descriptive terms appeared:

- typist's and cashier's wrist
- letter sorter's and yoga wrist
- golfer's or tennis elbow
- letter carrier's shoulder
- carpet layer's knee
- baseball catcher's hand
- typist's myalgia
- typist's tenosynovitis
- keyboard user's mouse elbow
- keyboard user's carpal tunnel syndrome
- nurses' back
- texter's thumb.

Repetition-induced injuries can also stem from activities related to leisure; for instance, knitting or playing golf and tennis. When related to work, however, they become ethical and legal issues. Fortunately, we now understand the biomechanical processes that can lead to overexertion and pathology of human tissue due to repetitive activities. With this knowledge, engineering as well as managerial interventions can avoid the occurrence of repetitive injuries. The ergonomics literature explains how to design tasks and tools to prevent musculoskeletal disorders (see, for example, Arndt & Putz-Anderson, 2001; Bernard, 1997; Carayon, Haims, & Smith, 1999; Karwowski 2001; Kroemer, 1997b, 2001; K. H. E. Kroemer & A. D. Kroemer, 2001a, 2001b; Kroemer et al., 2001; Kuorinka & Forcier, 1995; Moon & Sauter, 1996; National Research Council, 1999, 2001; Nordin, Andersson, & Pope, 1997; Putz-Anderson, 1988; Violante, Isolani, & Raffi, 2001).

Ergonomic solutions apply to practically all cases. Some are very simple, such as a decision by management to reduce time spent at repetitive tasks with vibrating tools. Although this organizational approach helps to avoid injury to employees, it does not solve the underlying problem of the work device or the job itself being of an injurious nature. The far better solution is to eliminate the cause by proper design or redesign of tools, processes, and tasks so that they do not pose any danger, as described by the aforementioned authors.

Low Back Problems

Back problems, especially in the lumbar area, are among the most common health complaints, and apparently have been since ancient times (Adams, Bogduk, Burton, & Dolan, 2002; Marras, 2000; Marras & Waters, 2004; National Research Council and Institute of Medicine, 2001; Riihimaeki, 2001; Snook, 2001). Low back pain is such an ordinary event that the abbreviation LBP now commonly suffices to describe this condition. LBP occurs at all ages, but incidence appears to be highest in the early teenage years, genetics being a strong factor (Leboeuf-Yde, 2004). Aside from acute injuries, disease, and age-related deterioration, the common cause of LBD is overuse, which strains the bones of the spinal column and its support tissues, ligaments, muscles, and tendons. These structures have inherent biomechanical weaknesses and usually deteriorate with age.

Low back injuries at work occur most frequently in the morning hours (Fathallah & Brogmus, 1997; Snook, Webster, & McGorry, 2002), possibly because at that time, the elastic tissues are more flexible than later during the day. Often internal and use factors combine to create strain as well as pain. The overuse event may be a single heavy effort or consist of repeated submaximal loadings. Even if each effort is, by itself, minor and of no discernible health consequence, numerous repetitions of these microtraumas can be harmful by their cumulative effects. Sedentary lifestyle does not appear to be a causal factor (Leboeuf-Yde, 2004).

Often the person is not aware that an injury has occured but will feel the effects later, commonly as stiffness and inability to execute certain movements. Pain may or may not be present, but if it does exist, it is a certain sign of a deficiency. An examination may reveal a preexisting injury that has not yet caused pain or even inconvenience to the individual. Medical treatment is necessary and useful in about only second case, and surgery should be reserved for the few patients in real need. In many other cases, pain and stiffness go away over time without specific treatment (Deyo & Weinstein, 2001).

LBP befalls many people, but often its source remains unknown, frequently its nature is difficult to diagnose, and it is time-consuming to treat. Most treatments are ineffective (Snook, 2004). Ergonomic help is available in two ways: by avoiding strainful body exertions by appropriate design of task and equipment and by using suitable body support structures (Silverstein & Clark, 2004). An individual's attitude, especially in terms of managing his or her work and doing physical exercise, is a very important factor as well.

Common sense tells us that a person trying to lift a load that is too heavy is at risk of a back injury. However, such a general statement has little biomechanical meaning and offers only gross practical advice because *any* load external to the body can impose high loading of body structures, depending on the load's mechanical advantage relative to the spine (National Research Council & Institute of Medicine, 2001, p. 236). Even just moving or twisting one's own body can lead to LBP. Thus, ergonomic design of the workplace (including tasks and tools) is not only a challenging undertaking (Kroemer, 1997b) but can also be truly effective. The best solution is to do away with the human effort either altogether by redesigning the process so that the handling of material is completely eliminated or taken over by some device.

Persons who already suffer from lower back pain but continue to attempt work that involves lifting and related back motions are especially at risk. Marras, Davis, Ferguson, Lucas, and Purnendu (2001) demonstrated that LBP sufferers, while attempting to avoid stressful exertions, actually may increase strain on the spine — though unintentionally — especially in compression and lateral shear. Excessive body weight, which is often associated with chronic low back pain, further increases spine loading (Marras et al., 2001). Weight control is of great importance to sufferers of LBP; naturally, they should keep loads small and do all lifting close to the trunk and in front of it.

TABLE 4.5
Approaches to Improved Matarial Handling

Task To Be Improved	Best Solution	Second-Best Solution	Partial Solution	Change Task to
Lifting and lowering	Eliminate task altogether or have all objects at working height	Increase load and then mechanize	Reduce load	Pushing and pulling
Carrying	Eliminate task altogether	Use belt, cart, or other equipment to move object	Reduce load	Pushing and pulling
Pushing and pulling	Eliminate task altogether	Use belt, cart, or other equipment to move object	Reduce load	

Adapted from Kroemer, K. H. E. (1997). *Ergonomic design of material handling systems.* Boca Raton, FL: CRC Press.

Today, it is generally acknowledged that many LBP sufferers are better off and heal faster when they continue to be active rather than staying in bed and resting, as used to be the medical advice (Snook, 2004). However, on the job this requires suitable equipment, tasks, and habits, as well as understanding supervisors.

Ergonomic design of the workplace helps people avoid back strain in the first place and enables those with back deficiencies to remain on the job or return to work soon after injury (Snook, 2001). Major activities to be avoided by design and training are

- trunk bending
- trunk twisting
- lifting, lowering, carrying, and pushing and pulling objects, even if light in weight.

These activities are especially dangerous if done in combination, and more so if often repeated. Kroemer (1997b) wrote in detail about ergonomic solutions to alleviate human material handling, including lifting. Among his suggestions is the scheme shown in Table 4.5 to avoid the most risky handling tasks at the workplace.

BEDS AND CHAIRS

It is interesting and even amusing to read about the many attempts to design beds (H. M. Parsons, 1972) that supposedly suit people with back pain. I, a longtime LBP sufferer, have observed how recommendations for bed surfaces changed from flat and hard sheets of plywood and rows of wooden slats in the 1960s to air- and water-filled mattresses in the 1980s; followed by foam mattresses and surfaces of varying thickness and consistency, with other high-tech support systems added later (Haex & van Houte, 2001b).

Over these decades, chair recommendations also changed, from hard and inflexible to soft and adjustable. Seat surfaces became flat or saddle-shaped and otherwise contoured, inclined or declined. Backrests were advised, denounced, and again advised. Rigidly upright backrests changed into angled and declinable back supports with adjustable lumbar pads. Some experts recommend standing rather than sitting, and others prefer sitting, if done "correctly." Current chair designs move with the body (K. H. E. Kroemer & A. D. Kroemer, 2001a; Kroemer et al., 2001; Tenner, 2003b).

The rocking chair belongs to a special group of seats that allow people with back pain to move their bodies slightly without losing support. President John F. Kennedy was probably the best-known user of rocking chairs (Fisher, 2001). Many of us who suffer from back pain have one comfortable chair that soothes us. We find it difficult to pinpoint the specific design feature, or set of features, that makes the chair so good; but being able to move the body while maintaining support seems to be especially important.

LIFT BELTS

Another approach to help people avoid back injuries is the use of *lift belts*, also often called *back belts*. These provide some sort of external wrapping around the abdominal region, meant to stiffen the walls of the pressure column in the belly. Among others, Lavender, Chen, Li, and Andersson (1998), McGill (1999), and Thoumier, Drape, Aymard, and Bedoiseau (1998) studied the effects of such belts. Their conclusions neither summarily support nor condemn the wearing of support belts in industrial jobs:

- Certain material handlers, especially persons who have suffered a back injury, may benefit from a suitable belt.
- Candidates for belt wearing should be screened for cardiovascular risks, which may be increased by belt pressure.
- Belt wearers should receive special training because the presence of the belt may provide a false sense of security.
- Belts should not be considered for long-term use.
- Belts are not a substitute for ergonomic design of the work task, work-place, and work equipment.

On the whole, the use of lift belts for professional material handling does not seem to be an effective way of preventing injuries caused by overexertion. Even competitive weight lifters, who usually wear such belts, suffer back injuries. Human engineered design of task and equipment, and suitable behavior are the keys to avoid LBP.

NURSES AND THEIR PATIENTS

Working as a caregiver is an occupation with high risk of low back injuries. Nurses and their aides must often lift and move patients — precious loads indeed, complex in shape, varying in size and weight, and with no handholds. They are often difficult to move because of limpness. Pain and discomfort can make them uncooperative and reluctant to be moved. But patients must be shifted between gurney and bed, lifted, and turned for changing gowns and bedding.

Lifting a limp, painful body is not only a torture to the patient but also a major physical effort for the care personnel. Many caregivers suffer musculoskeletal injuries, often of the back, when moving patients. Patient handling is associated with the highest rates of occupation-related back pain (Andersen, Schibye, & Skotte, 2001; Daynard, Yassi, Cooper, Tate, Norman, & Wells, 2001; Engkvist, Kjellberg, Wigaeus, Hagberg, Menckel, & Ekenvall, 2001; Theilmeier, Jordan, Jaeger, & Luttmann, 2003), exceeded only by material handling in warehousing (National Research Council & Institute of Medicine, 2001, p. 235). In terms of the number of incidents of nonfatal occupational injury and illness, the U.S. nursing home industry unfortunately ranks third among all U.S. industries. To improve the safety of caregivers, in 2003, the Occupational Safety and Health Administration (OSHA) issued a set of guidelines that target the nursing home industry.

Patients must be moved primarily in, from, or to beds. For this, caregivers may have to assume awkward positions, often bending low and reaching out, generating especially high loading on the spine and its support tissues. If one person alone cannot move the patient, it is advisable to call helpers. Unfortunately, the ability of teams of persons to lift is not simply the sum of the team members' individual lift capacities but considerably less. Furthermore, team lifting can lead to awkward postures that produce increased biomechanical strain in the lifters (National Research Council & Institute of Medicine, 2001, p. 246).

Moving patients is one of many jobs that involve lifting loads with high risk of physical harm to the body. One would expect that persons' learning from past experience and receiving instructions on how to do the job safely would improve their lifting skills and make overexertion less likely. Yet, after reviewing four decades of education on how to lift safely, Brown (1972, 1975) and Yu, Roht, Wise, Kilian, and Weir (1984) found no significant reduction in back injuries. In the 1980s, Snook (1991) compared the number of back injuries in companies that conducted training programs in safe lifting with the number of injuries in enterprises without such programs; again, there was no significant difference.

In 1997, Daltrov et al. published the details of a controlled study with nearly 4,000 postal employees. Experienced physical therapists trained 2,534 workers and 134 supervisors in a carefully planned and executed field study. During the observation period of more than 5 years, 360 persons reported back injuries, meaning the rate was 21.2 injuries per 1,000 worker-years of risk. The median time off work was 14 days. A comparison of intervention groups with control groups showed no effects of the education program: not on the rate of low back injuries, not on the cost per injury, not on the time off work per injury, and not on the rate of injury repetition. (Oddly, though, the specially trained employees showed an increased knowledge of safe behavior but became injured anyway.)

Perhaps most disappointing is the report by Scholey and Hair (1989), who found that 212 physical therapists, who themselves provided back care education, had the same incidence, prevalence, and recurrence of back pain as a carefully matched control group. However, contrasting these negative findings are those by Johnsson, Carlsson, and Lagerstroem (2002); Keir and MacDonell (2004); Schibye, Hansen, Hye-Knudsen, Essendrop, Boecher, and Skotte (2003); and Nussbaum, and Torres

(2001), who reviewed the literature on instructions to nurses on how to lift when handling patients, and observed patient transfers. Their encouraging findings are that some specific directives reduce the incidence of low back injury.

In general, it is still not clear whether, under which circumstances, and to what extent instructions on how to prevent back injuries are effective in patient handling or in general load handling. Some, possibly most, training approaches may be ineffective in injury prevention, or their effects so uncertain and inconsistent, that money and effort spent for training programs might be better used on ergonomic job design. "In spite of more than 50 years of concerted effort to diminish task demand, the incidence of compensable back injuries has not wavered. ... Rather than pursuing the 'right way to lift,' the more reasonable and humane quest might be for workplaces that are comfortable when we are well and accommodating when we are ill" (Hadler, 1997, p. 935, on industrial material handling).

Many better human engineered hospital beds, gurneys, and mechanical devices to lift and move patients have been proposed, but as a rule were found to be awkward and time consuming to use or simply too expensive (Elford, Straker, & Strauss, 2000; Gagnon, Akre, Chehade, Kemp, & Lortie, 1987; Le Bon & Forrester, 1997; Lynch & Freund, 2000; Marras, Davis, Kirking, & Bertsche, 1999; Ulin et al., 1997; Zhuang, Stobbe, Collins, Hsiao, & Hobbs, 2000). An efficient, safe, and inexpensive technique still eludes us. It may be necessary to devise new systems — radically different from traditional beds and gurneys — for supporting and moving patients to derive an overall better and truly human-engineered solution. Charney and Hudson (2004) discuss the underlying problems and new solutions.

SUMMARY

Given today's state of ergonomics knowledge and computer technology, it is inexcusable that some tasks, tools, and workstations are still designed for the phantom of the average person assuming a static posture. Instead, ranges of motion and variability in body size and strength must establish the design criteria.

Design for persons who are very small, very big, or who have other "extraordinary" body traits requires, in most cases, that their special characteristics be individually measured and not calculated from general tables.

Repetition-induced injuries have long been known to occur widely in certain occupations. They are now prominent among computer users. Current technology should be suitable to avoid the underlying causes.

Often, but not always, the engineer can "design out" low-back problems by well-chosen equipment, but mere instructions (for example, on how to lift objects) are usually not effective.

Transferring patients between gurney and bed and moving bedridden persons for changing clothes and bedding are among the urgent and so far unresolved human engineering tasks. Moving a limp, painful body is not only a torture to the patient but also a major physical effort for the caregivers, many of whom suffer musculoskeletal injuries, often of the back.

5 Design for Persons with Disabilities

OVERVIEW

Numerous kinds of disabilities make it difficult or even impossible to do activities that are normal to most people. Depending on how one defines *disability,* estimates of the percentage of afflicted working-age people in the United States range from about 8% to 17%.

The tasks of the ergonomist lie mainly in the area of preventing or at least alleviating disabilities by providing body assists through biomechanics (for example, as prostheses); by newly designing or redesigning processes, tasks, and procedures so that the affected person can perform them; and by devising engineered conditions (environment, equipment, and tools) to suit the person with an impairment. As a rule, the human factors engineer works with physicians and therapists to understand and properly consider the underlying impairments. In some cases, the engineer cooperates with medical specialists to counteract impairment directly, for example, by using nerve impulses to control an artificial organ.

Overcoming impairment — or at least reducing its impact — is possible in many cases, but this ergonomic process often involves sophisticated and expensive measures. Fortunately, increased public attention and greater consideration of the individual needs of persons with disabilities have led to both improved assessment techniques and more efficient assistive technology.

DEFINING AND MEASURING DISABILITIES

Even normative regular adults are only temporarily able-bodied: As children, we lack the strength and skill that we will acquire later; while aging, we lose capabilities that we previously enjoyed. Injury or illness can and does deprive us of certain abilities, perhaps only for a limited time, but possibly forever.

In 1978, a special issue (Volume 20, Number 1) of the journal *Human Factors* specifically addressed persons with impairments. In the introduction, the guest editor deplored that remarkably few studies dealing with problems of "the handicapped" had appeared in print until then. The next such special issue of *Human Factors* appeared in 1990 (Volume 32, Number 4), titled "Assisting People with Functional Impairments." Its guest editor stated that full-time employment of human factors specialists in the rehabilitation field was still very rare.

Fortunately, by now the situation has substantially improved. The concern of ergonomists and human factors engineers for the welfare of persons with disabilities, interest in their rehabilitation, and even "reengineering" the biomechanically or physiologically impaired body have become widespread in recent years: for examples, see publications by Elkind, Nickerson, Van Cott, and Williges (1995); Gardner-Bonneau and Gosbee (1997); Jensen (1999); Kroemer and Grandjean (1997); Kroemer et al. (2001); Mital and Karwowski (1988); Rice (1998); Vanderheiden (1997).

It is not only a humanitarian goal to manufacture equipment suitable for persons with impairments but also a challenge to business. That challenge has two aspects: One is the marketing opportunity of selling to the many potential buyers; the other is the problem of the lack of data that describe the functional capabilities or disabilities of these prospective customers. This lack of a comprehensive descriptive database is evident in Europe and elsewhere across the globe (Feeney, Summerskill, Porter, & Freer, 2000). Yet, using the limited systematic data available, CAD (computerized automated design) systems have proven to be highly effective tools to devise wheelchairs and powered vehicles, workstations and habitats, and devices and tools, as Feeney et al. (2000) and Greil and Juergens (2000) described.

The Americans with Disabilities Act (ADA), signed into law in 1990, contains this definition for *disability:* "a physical or mental impairment which substantially limits one or more of an individual's major activities of daily living such as walking, hearing, speaking, learning, and performing manual tasks." The Committee on an Aging Society of the U.S. National Research Council (1988) defined disability, as did the World Health Organization, as "any restriction or lack (resulting from an impairment) of ability to perform an activity in the manner, or in the range, considered normal."

Using these classifications and those proposed by Levine, Zitter, and Ingram (1990) and Verbrugge (1991), we arrive at the following definitions that I use in this book:

- *Impairment* is a physiological, psychological, or anatomical abnormality of body structure or function. Major categories of dysfunctions are in sensory, motor, and cognitive domains. The impairment may be chronic (long term, even permanent) or temporary. The impairment often causes a *disability* to perform certain activities.
- *Disability* means that the task is impossible to do as required or that it takes "extra-ordinary" effort and unusual time to perform the action.
- *Handicap* is the physical, social, and economic disadvantage that results from impairment or disability. It often entails loss of social status, contacts, and income.

Even with these widely used definitions, judgments vary broadly regarding impairments, disabilities, and handicaps with respect to both their nature and their extent.

PERFORMANCE IMPAIRMENTS AND DISABILITIES

Performance impairments and disabilities may result from very dissimilar conditions; for instance, poor balance can stem from conditions related to blood pressure,

TABLE 5.1
Common Impairments in Everyday Tasks Associated With Underlying Individual Disabilities

Task Impairments	Individual Disabilities
Difficulty in interpreting information	Reduced abilities to read or reason
Difficulty in interpreting visual information, knowing where one is, finding one's way	Legal blindness (20/200), cannot read ordinary newspaper print even with eye correction lenses, vision-field defects of 10% or more
Cannot understand usual speech with or without amplification	Severe loss of hearing
Limited to looking up and down, or sideways	Difficulty in moving the head
Cannot grasp, place, or direct objects	Lack of coordination in controlling the movements of the extremities
Difficulty in lifting arms, reaching, and performing forceful manipulation	Decreased range of motion and strength of the upper extremities
Difficulty in performing manipulation of objects	Decreased strength, mobility, and control of the hand and its digits

Adapted from a listing in *Access to the Built Environment: A Review of Literature* (PB80-136526), by Steinfeld et al., 1979, Washington, DC: Superintendent of Documents.

hemiplegia, amputation, muscular dystrophy, Parkinson's disease, brain tumor, and other causes. To remedy or alleviate the outcome of poor balance, very different medical or ergonomic treatments would have to be taken, depending on the underlying deficiency.

Table 5.1 associates underlying individual disabilities with resulting common impairments in everyday tasks. Often, an individual suffers from several impairments that can combine to a severe disability.

Decades ago, Faste (1977) and Steinfeld, Schroeder, Duncan, Wirth, and Faste (1979) showed that it can be useful to assemble major human faculties into groups; for example, concerning information, coordination, power, manipulation, and ambulation. Such groupings help to systematize the ergonomist's task. As Figure 5.1 indicates, the information group can be sorted into components received through the primary senses: feeling (taction), seeing (vision), and hearing (audition) and the processing of that information. Generating sounds, especially words, and even eye movements convey information to others. (Chapter 1 contains a detailed discussion of some of these faculties, and Chapter 2 indicates methods for their assessment.)

Impairments in a person's output capabilities (human *efference*, feedforward) and input capacities (*afference*, information including feedback to the human) diminish the individual's ability to perform certain tasks. (The direction of arrows in Figure 5.1 indicates efferences and afferences.) In cases of impaired information processing in the brain and of reduced vision, both efferent and afferent functions are diminished because vision can be used as output by "tracking the eye." Diminished hearing and feeling (temperature, pain, taction, and kinesthetics) primarily

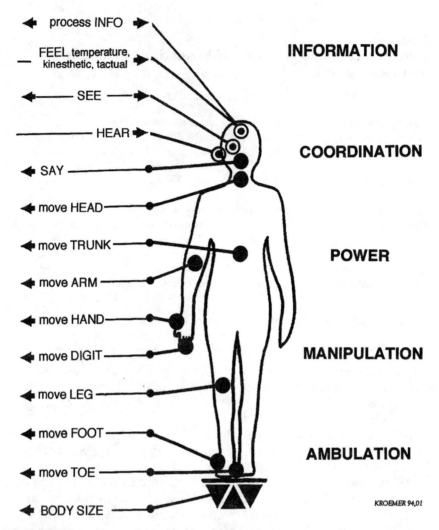

FIGURE 5.1 Categorization of human capabilities for information exchange, body coordination, power, manipulation, and ambulation.

decrease input capabilities, whereas the other impairments primarily reduce output capabilities.

A person's abilities in body coordination, power, manipulation, and ambulating serve both to move one's own body and to operate controls, tools, and machinery of all kinds. These capabilities are difficult to classify into single channels because they interact among themselves and with information; but generally speaking, they are the abilities to move the head, trunk, arm, hand, digit, leg, foot, and toe. Each on these can be measured (see Chapter 2) and so can the size of the body, which, especially if extreme in length or weight, can severely affect motion capabilities.

PROBLEM IDENTIFICATION MATRIX

Such organization of human capabilities into groups allows establishing *problem identification matrices* (PIMs). They show systems requirements on one axis and human capabilities (and especially limitations therein) on an orthogonal axis. At the intersections of these axes, symbols indicate discrepancies. These symbols show which features of a given design (say, of a public announcement system or of controls in a truck) pose problems for persons with certain types of impairments. This knowledge, in turn, makes it easy to identify the specific ergonomic means to enable groups of persons who are afflicted with certain classes of impairments. Likewise, this matrix-based information helps to determine how to assist any given person who needs special attention. Figures 5.2, 5.3, and 5.4 illustrate specific PIMs for three distinctive applications: a communication system, controls at a workstation, and plumbing fixtures in a bathroom, respectively.

The symbols placed at the points where system requirements and human capabilities intersect can indicate the severity of the problem and hence the priority for ergonomic improvements.

After having pinpointed, for instance via PIMs, the difficulties encountered by persons with impairments while attempting to do a task, ergonomic countermeasures become apparent that, when taken, enable or improve task performance. PIMs appear especially effective for the purpose of convincing administrators and politicians that generic design features — say, stairs — make use of certain designed commodities — say, subways — impossible for certain groups of people with impairments, such as wheelchair users. Conversely, PIMs indicate which properly human engineered systems facilitate use by persons with impairments; examples are so-called kneeling buses with a flat floor and no steps and low sidewalk curbs at pedestrian crosswalks. PIMs also help to plan products well within the concept phase rather than redesigning them later; for example, by installing elevator platforms into existing buses or by installing low curbs at street crossings, which help wheelchair users.

"ERGONOMICS FOR ONE"

Accommodating large groups of people is the most common task of the human factors engineer. Examples are designing hand tools for assembly workers, developing night vision devices for soldiers and police, and improving keyboards or other computer input devices for normal people. Although smaller in group size than the general public, there are clusters of persons with impairments who all have the same disability, such as blindness or paralysis of the legs. Accordingly, the human factors engineer can develop and apply general ergonomic solutions — for instance Braille or wheelchairs — that help all these persons.

However, many disabilities are specific to an individual, especially if several exigencies combine, and that person's activity needs may be unique, as might be the performance aspirations. A single injury — for instance, to the head — may bring about various impairments in vision, memory, and motion control. The impairments

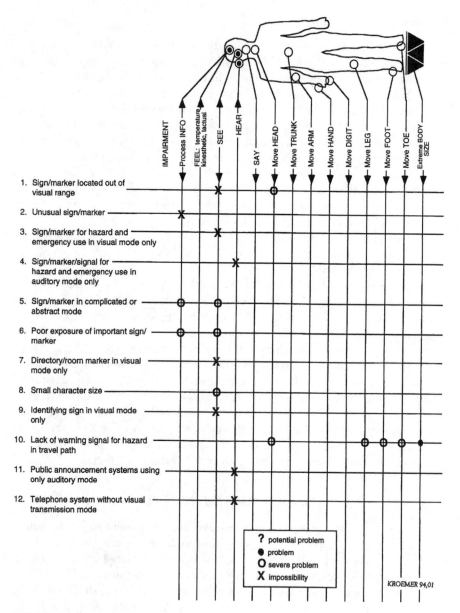

FIGURE 5.2 Communications PIM.

experienced by other victims of such a head trauma may be different. Furthermore, persons — for example, a carpenter, an architect, or a telemarketer — with similar afflictions may have distinct performance needs. Two persons with multiple sclerosis may have lost the use of their legs and therefore need wheelchairs for moving about, yet one person may be able to control and propel the wheelchair with the arms,

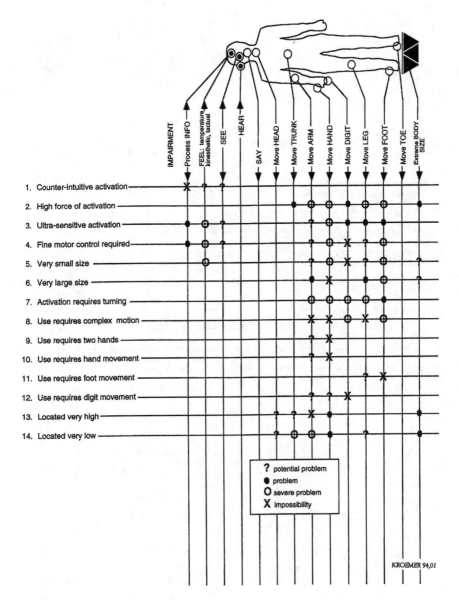

FIGURE 5.3 Controls PIM.

whereas the other may have also lost upper-body functions and may need a powered chair with complex controllers.

Instead of addressing the needs of a group of persons who share the same disabilities (Bradtmiller, 2000; A. M. Koontz et al., 2004), in many cases the task of the ergonomist is to determine particular designs that can help one individual with very specific impairments. This includes first ascertaining the person's individual capabilities and limitations and then providing specific assistance by removing

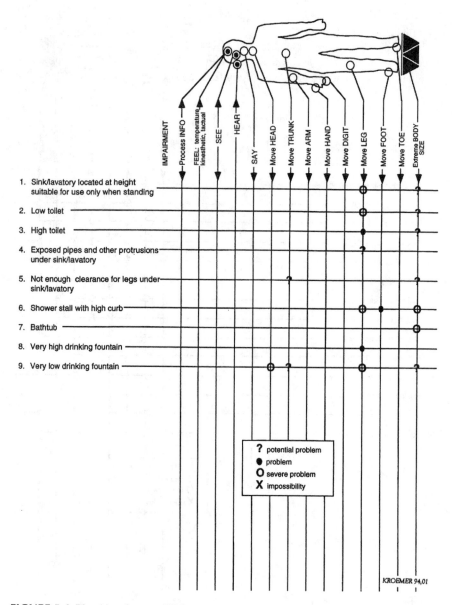

FIGURE 5.4 Plumbing fixtures PIM.

barriers and modifying the person's environment including equipment features and task requirements, thus augmenting performance and well-being. McQuistion (1993a) called this individual-centered work *"designing for one."*

Many people with disabilities have undergone medical evaluations, but the result rarely provides the specific information that the ergonomist needs to select and implement means to overcome the existing impairments. Methods and techniques described in Chapter 2 can supply that information. Measurements of body size, space requirements, and reach capabilities require special equipment and procedures (Feathers, Polzin, Paquet, Lenker, & Steinfeld, 2001; Steinfeld, 2001). To measure residual capabilities that persons with impairments may have for manipulation and for performance of defined tasks, Kondraske (1988); A. B. Swanson, Goran-Hagert, and G. D. G. Swanson (1987); and Triolo, Reilley, Freedman, and Bertz (1993) developed special testing devices and procedures.

In the 1970s, the Rehabilitation Engineering Center of Wichita State University was one of the first institutions to systematically and successfully apply ergonomic solutions to assist persons with severe physical impairments (Bazar, 1978). A case in point is the Available Motions Inventory (AMI), described by Dryden and Kemmerling (1990) and R. V. Smith and Leslie (1990). The AMI used a set of panels that contain switches, knobs, and other devices to be reached, activated, and manipulated. These panels were put into standardized positions within the work area, and a person's ability or inability to operate the controls were recorded together with time needed and strength exerted, as appropriate. A large available database allowed the comparison of the individual's performance with that of others, whether disabled or not. Thus, the individual capability or limitation for certain tasks of particular interest could be determined. Such information facilitates a systematic approach to finding work tasks and work conditions suitable for a given individual's capabilities or to modifying them to match the person's abilities.

The now-classic *Tech Briefs* of the Rehabilitation Engineering Center of Wichita State University describe a large number of cases in which individuals with impairments were enabled to perform tasks, even in gainful employment, through the use of specially engineered devices, such as

- cap counter, lift table, marine chair, scanner, measuring device, material handler (all in *Tech Briefs* 1/1993)
- lower-body pressure suit, voice amplifier (2/1993)
- reclined and supine computer workstations (3/1993)
- children's shoelace-tying aid (2/1994)
- automatic inhalation device (4/1994)
- wire-winding machine (1/1995)
- piano-tuning device (3/1996)
- toileting aid (2/1996).

The literature demonstrates the many possibilities of applying human factors to enable persons with impairments to function. For example, McQuistion (1993b) described

- a speaking Braille machine
- custom wrist supports
- a rotating palette leveler.

As the work done at Wichita State University and other institutions demonstrates, nearly every specific impairment can be overcome, at least partially, by ergonomic means, though often at high cost and effort. To inquire about the specialties and locations of private and public institutions that can do such work, see the Web sites listed at the end of this chapter.

LOCOMOTION AIDS

Locomotion and transportation are serious problems for many people with disabilities. Providing aids in the form of crutches, walkers, and prostheses has been a human engineering task since ancient times. Powered wheelchairs and automobiles are newer solutions.

WHEELCHAIRS

Past designs of wheelchairs for those who cannot walk have been described by, among others, Bloswick, Shirley, and King (1998); Gilad (2001); Kamentz (1969); Troy, Cooper, Robertson, and Grey (1995); Van der Woude, Meijs, and de Boer (1991); and Zacharkow (1984, 1988). Special features can include ramp and stair climbing, use in showers or toilets, devices for transfer to and from the chair, and provisions to elevate and lower the user. There are special wheelchairs for indoor and outdoor use, for fitness training, and even for racing. Various kinds of propulsion may be generated and controlled either by the seated client or by an attendant or provided by an auxiliary source, commonly an electric or combustion engine.

Wheelchairs not only enable mobility, they must also support the trunk and the legs, allow the seated person to manipulate objects in front and at the sides, be stable, be lockable in place, and facilitate sitting down and getting out of the chair. Specific design features of seat pan, backrest, armrest, or leg support can meet the need to provide stable yet relaxed body support, at the same time avoiding pressure points and resulting sores (Troy et al., 1995; Zacharkow, 1984).

Wheelchairs require passageways that are flatter and wider than a walking person needs. This demand can be easily met in newly planned environments (W. L. Wilkoff, P. C. Wilkoff, & Abed, 1998), outdoors or indoors, but modifications of existing facilities can be costly and may be difficult, especially inside buildings where the available space is limited. Therefore, early implementation of ergonomic measures, best during the planning stage, is highly desirable.

AUTOMOBILES

For many decades, human factors engineers have directed their efforts at making it possible or at least easier for drivers or passengers with disabilities to use automobiles (Koppa, 1990; Roush & Koppa, 1992). These improvements mainly concerned seating, hand and foot controls (often exchanging one for the other), ingress and egress, and the ability of the driver to see instruments and events on the outside. Today, a large variety of adaptive devices are available to help individuals with impairments operate virtually all the driving controls of automobiles (see, for example, the Society of Automotive Engineers [SAE] recommended practices and standards J1139, J1725, J1903, J2092, J2093, and J2094). It is relatively easy to reassign, extend, and otherwise modify vehicle controls to allow persons with impairments to steer, brake, and accelerate: these are the so-called Group A primary driver controls. Group B includes the accessory controls that operate the remaining driving functions such as turn signals and windshield wipers. If improperly designed, these secondary controls can cause distraction and even interference with the use of Group A primary controls and so degrade the driver's performance.

Priority ranking of the secondary controls is one of the first steps in making decisions regarding their design and location. Table 5.2 shows, as an example, rankings made by 32 members of SAE's Adaptive Devices Standards Committee. These are rankings pooled from the experts' answers to the following questions:

While the vehicle is in motion, the driver

- MUST operate this control *as soon as possible*?
- SHOULD operate this control *as soon as convenient*?
- MAY or MAY NOT operate this control?
- does NOT NEED to operate this control?

TABLE 5.2
Ranking of Secondary Automobile Controls

Control	Mode	Mean	Standard Deviation
Horn	1	1.22	0.61
Turn signal	1	1.25	0.44
Headlights — dim	1	1.42	0.67
Windshield wiper — low speed	1	1.62	0.75
Windshield wiper — high speed	2	1.66	0.60
Windshield washer	2	1.78	0.79
Headlights — on/off	2	1.91	0.78
Cruise control — set	2	2.03	0.74
HVAC — select mode	2	2.16	0.85
Cruise control — on/off	2	2.19	0.96
Windshield wiper — single swipe	3	2.25	1.02
HVAC — temperature	2	2.34	0.79
Power windows	2	2.41	0.76
Automatic transmission mode	4	2.55	1.21
Hazard warning	2	2.56	0.91

Modified from "A Survey of Activation Importance of Individual Secondary Controls in Modified Vehicles," by L. Roush and R. Koppa, 1992, *Assist Technology, 4*, 66–69.

Note the differences in opinion discernible from the differences between mode and mean values in Table 5.2 and from the magnitudes of the standard deviations. This table is an excerpt from the original 1992 compilation that included 23 controls typical of the state of technology at that time. It would be interesting to see how such rankings change with newer automotive designs, traffic patterns, and driver behaviors.

Several periodicals provide information about recent developments, among them the journals *Biomechanical Engineering, Disability and Rehabilitation, IEEE Transactions on Rehabilitation, Prosthetics and Orthotics International, Rehabilitation R&D Progress Reports,* and *Rehabilitation and Research Development.*

USE OF COMPUTERS

Computers have three special traits that make them particularly helpful for many persons with disabilities. Computers are

- relatively easy to use as an input or output device
- capable of processing complex information rapidly
- able to control other equipment.

Yet, the input interface to the computer particularly requires ergonomic attention. The customary QWERTY keyboard is basically user unfriendly (Helander, Landauer, & Prabhu, 1997; Marklin & Simoneau, 2004; Kroemer, 2001; K. H. E. Kroemer & A. D. Kroemer, 2001a; K. H. E. Kroemer, H. B. Kroemer, & Kroemer-Elbert, 2001; Treaster & Marras, 2000; Tenner, 2003b), and its long-known faults warrant a fundamental redesign. However, one does not have to be able to write by hand, draw, or manipulate paper files to use a keyboard, even though it is ergonomically deficient. The current multikey QWERTY keyboard or an ordinary mouse may be difficult or impossible to use for a person who has little or no control over hand and arm movements. Thus, adaptive hardware is often needed, possibly as simple as a stick attached to the head or hand to press the keys. Other devices are more complex, such as a sensor that recognizes the gaze of the eye or certain body motions and activates an associated key (Calhoun, 2001; McMillan, 2001; McMillan & Calhoun, 2001).

An individual with severe muscular disabilities such as cerebral palsy or multiple sclerosis may find a regular keyboard impossible to use even with a stick and thus would need a special input device. A theoretically appealing solution is speech recognition software. However, in spite of many years of development work, some commercially available programs are still not easy to use because of imperfections in terms of correct interpretation of spoken inputs and spelling. Voice output can speak computer text that simultaneously appears on the monitor screen.

The location of the common visual display needs careful consideration to suit the user: As K. H. E. Kroemer and A. D. Kroemer (2001a) and Sommerich, Joines, and Psihogios (2001) explained, even for users with normal vision and healthy spines, the screen often is too far away and too high. The frequently heard but false admonition to place the monitor at eye height is a relic from the late 1950s, when human factors aspects of consoles were of great concern, especially in power plants. These consoles had a large number of hand-operated controls and a great variety of associated dials. With priority given to the location of the cranks and wheels, levers and knobs, switches and buttons, the available console space from waist to shoulder height was assigned to them. This left only the area above, at about eye height, to place the displays. Today's computers are different from those consoles in terms of control input and display output. Whether desktop or laptop computer, we can place the keyboard and mouse at elbow height, and locating the display low and directly behind the keyboard makes close viewing naturally easy.

Many software programs are available that have been specifically developed for individuals with disabilities. Some go far beyond common computer use; they may even include the control of the environmental conditions in a home or office. Casali and Williges (1990) described a systematic process to determine the most appropriate aid for a computer user with a disability. The approach includes three basic steps. The first task is to assess the residual abilities of the client regarding computer use. Then follows the search for a special database containing information about available hardware and software from which one selects a candidate solution. The final step is to test the usability and usefulness of the chosen solution and make changes and improvements as needed (August & Weiss, 1992).

Computer technology greatly facilitates impaired persons' abilities to communicate with other people, which is especially important if one cannot physically ambulate to another location or travel to an off-site meeting. R. C. Williges and B. H. Williges (1995) coined the term *surrogate travel* to indicate that only the information travels, not the person. Conferencing technology can be used for establishing remote work facilities (such as a home office), shared work, training, and other electronic linking such as in satellite conferencing. However, before the electronic "virtual workspace" as a mix of audio, video, and computer links becomes a reality, a number of issues need to be resolved.

As the Willigeses explained, a comprehensive taxonomy of electronic office tasks must be developed to determine proper information management concerning purpose and task (what is to be done in general and how so specifically) as well as conditions (informal, controversial, etc.). These overall categories need modification according to type and size of the meeting, ranging from multiperson conferences through managerial meetings, group work, dialogue among just a few or two persons, and single-person use. Each category may require hardware and software of different levels of complexity, with the lowest grade needed for a designated single user.

Developments in computer technology have opened up the Internet and thereby made available an abundance of information to people who otherwise would be isolated because of illness, injury, other disability, or age. Unfortunately, many potential users encounter problems associated with the physical use of the technical equipment itself (see the preceding text) because system designers do not seem to consider sufficiently that persons with impairments are active common users: Ergonomic deficiencies in keyboards, other input devices, or manuals and instructions are particularly annoying and hinder efficient operation. Search, choice, and integration of information have created large volumes of information and hence new tasks of aggregation and use of that information (Czaja, 1997, 1998). However, in spite of remaining deficiencies, that new technology has opened the door for many persons who would otherwise be shut off from communication with others to work, shop, or bank.

Computer technology can also be of great help to caregivers. It can link them to their patients and fellow caregivers, supervisors, and medical staff. It provides caregivers with access to online assistance programs and other databases, but the quality of these and their possible legal ramifications are currently unresolved issues (Czaja, 2001a, 2001b). Furthermore, proper use of computer systems does impose knowledge and skill requirements on the caregivers that for at least some of them might tax their capabilities.

ERGONOMIC DESIGN OF TELESUPPORT SYSTEMS

The term *telesupport* describes computer-based assistive technology that electronically delivers such services as health care advice, vocational rehabilitation training, social support, community networking, and other information of particular importance to shut-ins, sick or injured persons, and people with disabilities. Until recently, development of such telesupport systems has focused more on technical and content features than on the demands of their users for usefulness and ease of use. This has led to common complaints about product complexity and difficult and frustrating usage requirements. Such experiences often make users avoid the product or service from the beginning or cause them to eventually discontinue its use; both results hurt the vendor and the customer.

In the user-centered approach, the ergonomist starts by assessing the consumers' needs for type, content, and form of service — and their use-relevant capabilities and limitations. Basing the further development of service or product on these assessments avoids the risk that users with disabilities and some older people, as well as ethnic or social class minorities, will become marginalized or kept outside the system instead of being its beneficiaries as originally intended. Ergonomic design of telesupport systems, as a rule, not only helps disadvantaged persons but also makes the use of the services easier for all.

Human-engineered design best begins with careful and exhaustive compilations of use and users' expectations, requirements, and specifications. Based on this fundamental knowledge, prototypes can be designed and evaluated and then redesigned as needed until consumer satisfaction is achieved. This iterative process must be done in carefully planned usage tests that involve not only ergonomics experts but also, and importantly, actual users. The usage tests can be done in either laboratory focus groups or the users' everyday environment. Such participatory design has proven to lead to solutions that are practical, convenient, and accepted by the consumer (W. A. Rogers & Fisk, 2001; Scherer & Lane, 1997).

The traditional methods of data collection, first to assemble basic information about the prospective user and then information about the usability of a prototype product, were of the face-to-face type in which researchers physically visited the consumers. Although there are advantages to this approach, today much of this work is done online in computerized use- and user-centered ethnographic methods. The term *ethnographic*, borrowed from anthropology, indicates that qualitative information and subjective experiences are sought in addition to quantitative data. A typical example is to conduct usability tests in the users' daily environment, not in a laboratory. That is likely to yield more realistic information than inferences drawn from performance in an artificial laboratory situation.

With today's technology, much information about users and their habits, and especially about actual use of a prototype or operational product, can be obtained electronically. Smith-Jackson and Williges (2001) discussed such methodologies, including remote needs collection tools and on-line focus groups.

Ergonomic design necessary to accommodate persons with impairments usually makes work and life easier for everybody, including the able-bodied. Because becoming old often means losing capabilities that one once possessed, many of the human factors solutions that help the aging (discussed in Chapter 6) are also helpful to persons with disabilities, and vice versa.

CONSUMER PRODUCTS

Many everyday consumer products such as computer keyboards, household utensils, remote controls, "child-proof" medicine bottles, door openers, vacuum cleaners, stoves, razors, faucets, and audiovisual equipment are difficult to employ even for most adults but impossible to use for a person with an impairment. Often, relatively small changes in the selection of the control type, direction, and effort of operation (Daams, 2001; Freivalds, 1999) can make the use of a product feasible for the person with an impairment. This improvement usually makes its use easier and safer for everybody: Many examples may be found in the work of Fisk and Rogers (1997); H. E. Hancock, Fisk, and Rogers (2001); Karwowski (2001); Karwowski and Marras (1999); Kroemer et al. (2001); and Salvendy (1997).

SELECTION OF ASSISTIVE TECHNOLOGY

The use of ergonomics knowledge in rehabilitation engineering is diverse, ranging from wrist splints to artificial limbs, from walking aids to special automobiles. Technology for people with disabilities has advanced beyond gadgeteering and now approaches systematic development and implementation of sophisticated devices. Provision of artificial knee, hip, and finger joints helps many persons who injured their joints or who suffer from arthritic diseases; in the United States, about half a million artificial knee and hip joints are implanted annually. Electromyogram (EMG) signals, movements of unimpaired body parts (such as the jaw), voice control, or direction of gaze can control motions of prostheses.

A variety of criteria exist to select such assistive devices (Batavia & Hammer, 1990; Fernie, 1997). The selection criteria include

- effectiveness — the extent to which the device improves the user's capabilities, independence, and the user's objective and subjective situation
- affordability — the extent to which the purchase, maintenance, and repair cause financial hardship to the user

- dependability and durability — the extent to which the device operates with repeatable and predictable accuracy for extended periods
- physical security — the probability that the device will not cause physical harm to the user or other people
- portability — the extent to which the device can be readily transported to different locations and operated there
- usability — the extent to which the consumer can easily learn the usage of a newly received device and can use it easily, safely, and independently for the intended purpose
- physical comfort and personal acceptability — the degree to which the device provides comfort, or at least mitigates pain or discomfort to the user, so that it attracts usage in public or private
- flexibility and compatibility — the extent to which the device can be augmented by options and to which it will interface with other devices used currently or in the future.

SOURCES OF INFORMATION AND ADVICE

The U.S. Census Bureau Web site (http://www.census.gov/hhes/www/disability.html) has a section on disability statistics. The National Center for Health Statistics provides information gathered in the National Health Interview Survey on Disability 1994 and 1995 data (http://www.cdc.gov/nchs/about/major/nhis_dis/nhis_dis.htm).

Many universities have departments that can provide information and help regarding ergonomics resources to assist a person who wants to overcome a disability; especially departments of architecture, engineering (particularly human factors/ industrial and systems engineering), gerontology, psychology, and physiology may be able to provide specialized advice. Examples of such resources are the Center for Rehabilitation Technology at the Georgia Institute of Technology (http://www.gatech.edu), the Center for Universal Design at North Carolina State University (http://www.design.ncsu.edu/cud), and the Trace R&D Center at the University of Wisconsin (http://www.trace.wisc.edu).

One might also approach specialized entities such as the American Occupational Therapy Association (http://www.aota.org) and the National Institute on Disability and Rehabilitation Research (http://www.abledata.com). InfoUse (http://www.infouse.com/) has an "Access to Disability Data" project funded by the National Institute on Disability and Rehabilitation Research and provides up-to-date listings of organizations active in disability- and accessibility-related areas. Local and regional agencies and government offices can be helpful as well.

The following list relies on compilations by the Trace R&D Center (personal communication, R. Lueder, January 27, 2003) and contains Web sites that, in turn, contain links to many other useful sites:

- Guide to Disability Resources on the Internet (http://www.disabilityresources.org/)
- National Organization on Disability (http://www.nod.org/)
- World Institute on Disability (http://www.wid.org/)

- The American Foundation for the Blind, or AFB (http://www.afb.org/)
- The American Council of the Blind, or ACB (http://www.acb.org/)
- The National Foundation of the Blind, or NFB (http://www.nfb.org)
- Lighthouse International (http://www.lighthouse.org/)
- National Association of the Deaf, or NAD (http://www.nad.org/)
- Self Help for Hard of Hearing People, or SHHH (http://www.shhh.org/)
- TDI, formerly Telecommunications for the Deaf (http://www.tdi-online. org/)
- The Disability Statistics Center (http://www.dsc.ucsf.edu/)

SUMMARY

Disability is a physical or mental impairment that limits an individual's capacity to perform major activities of daily living. The ergonomics emphasis is often on helping an individual, rather than groups of persons, because most impairments vary from person to person and may change a great deal with time. The impairment may have different sources and may result in diverse consequences, depending on the person's circumstances and aspirations.

A variety of special testing devices and procedures have been developed to measure residual capabilities, and an ergonomics subspecialty has developed, called *rehabilitation engineering.* Yet, the measurement of individual capabilities is often a costly and time-consuming process, not easily available to every person with a disability; and the means that can help to improve the person's abilities are often complicated and costly. Various mechanical assistive devices are commercially available, with wheelchairs among the best known; computers are often especially useful because they can be operated with little physical effort and can provide complex "brainlike" functions.

Overcoming a disability, or at least reducing its impact, is possible in many cases, but this ergonomic process may involve sophisticated and expensive measures. Fortunately, increased public attention and consideration of the individual needs of persons with disabilities lead to both improved assessment techniques and more efficient assistive technology.

6 Design for Aging

OVERVIEW

We can expect to work and live longer than our ancestors did — but, as they experienced, we still will face declines in our capabilities and functions as we get old. Which traits decline, and how fast and how much they decrease, will vary with age and among individuals.

Using chronological age as a time scale, researchers find an immense variety of maintained (perhaps, in some respects, even increased) faculties, decreasing capabilities, and either slow or fast changes with advancing age. As the numbers of aging persons and their percentage in the overall population become larger, the need for complete, reliable, quantitative, and valid information about this population swells.

A common assumption is that older people find high-technology or high-demand activities (such as piloting, air traffic controlling, driving, working overtime, and doing shift work) more difficult to perform than younger persons and that this difficulty increases as one ages. Yet, in familiar activities, age-related performance decrements may be small or unimportant: "Experience prevents decline," one may hope, and perhaps one's own capacities will remain on a high level into a ripe age.

This chapter presents an overview of these changes and the ergonomic means to design our environment so that we can function and live well while aging.

WHAT TO EXPECT AS ONE AGES

DEMOGRAPHICS

Demographic information shows that in North America, Europe, and in many other areas, the aging portion of the population has increased recently in both absolute numbers and percentages. Older employees now constitute a large segment of the workforce (Annis, Case, Clauser, & Bradtmiller, 1991; Coleman, 1993; Czaja, 1990, 2001b; Marquie, Cau-Bareille, & Volkoff, 1998; Panek, 1997; Serow & Sly, 1988; Soldo & Longino, 1988). In the United States in March 2002, more than 13% (nearly 4.5 million) of those 65 years and older (almost 34 million) were still working, according to the 2003 U.S. Census Bureau report. Of those, nearly twice as many were men than women. Here are the probable reasons for the significant increase, since the early 1990s, in the number of older persons still in the workforce:

- simply liking to do work
- enjoying longer and healthier lives
- returning to work after raising children.

Most likely, many also continue to work for economic reasons: Of those 65 and older, 10% live in poverty as defined by the U.S. Census Bureau, and many find their incomes smaller and health care expenses larger than they expected.

The National Research Council & Institute of Medicine (2001, p. 336) predicted that around the year 2010, persons 55 years and older will comprise about 20% of the entire U.S. workforce. S. Smith, Norris, and Peebles (2000) estimated that, worldwide, the number of people older than 65 years will almost double so that, by 2025, about every 10th person will be at least 65 years old. For the United States, in 2003, the Centers for Disease Control and Prevention (CDC) predicted that by 2030, Americans 65 years and older would number 71 million, nearly 20% of the entire population.

TERMINOLOGY

No definite boundaries exist that determine when a young person turns middle-aged and then elderly or old. According to a curious use of terms in the United States, a middle-aged person becomes an older person at 40 or 45 years of age, then elderly at about 65 years, old at 75 years, and very old (or even old-old) if one lives beyond 85 years. Fisk et al. (2004) call persons aged from about 60 years to 75 *younger-old* and those older than 75 years *older-old.*

LIFE EXPECTANCY

How long a person actually lives depends on diseases, accidents, wars, climate, hygiene, nutrition, gender, and genetic heritage. Approximately 3,000 years ago, the average human life span was about 20 years. It reached only the low 20s in ancient Greece and Rome. In the Middle Ages, around a thousand years ago, life expectancy in Western Europe had increased to the 30s. It was only in the low 40s at the end of the 19th century. In Colonial America, the average life span was merely about 35 years around 1700. In the following two centuries, it increased substantially to nearly 50 years by 1900 (Committee on an Aging Society, 1988). Within the last century, the average life span grew even further; according to 2005 data from the U.S. Census Bureau, in 2003 it was 80.1 years for females and 74.8 years for males.

HOW WE AGE

The human body ages in many noticeable ways. By about 40 years, one is likely to notice that sight and hearing begin to decline, as do many internal functions. Body fat starts to collect in the abdominal area, making skin elsewhere wrinkle. Of the two leading theories of aging, one tries to explain the process as a senescence that is programmed by genes. The other says that cells suffer cumulative damage from free radicals, toxins produced by their own internal chemical reactions.

Obviously and apparently inevitably, the human body and mind age, losing more and more of the capabilities that were still available in the middle years (Belsky, 1990; Birren & Schaie, 2001; Panek, 1997; Spence, 1994; Steenbekkers & Beijsterveldt, 1998). Nayak (1995) provided the following breakdown of functional impairments reported by 3,752 persons 55 years or older who lived independently:

locomotion	35%
bending	28%
remembering	22%
twisting	20%
reaching	18%
hearing	13%
grasping	13%
seeing	11%
eating	5%

Kramarow, Lentzner, Rooks, Weeks, and Saydah (1999) reported the results of interviews with noninstitutionalized persons in the United States that were conducted in the mid-1990s. Among those 70 to 74 years old, about every second person felt that she or he had difficulties in performing activities of daily living, and the problems increased sharply with getting older. Joint arthritis was by far the most frequent cause limiting the respondents' activities.

"SUCCESSFUL AGING"

In the positive view of the ergonomics engineer, most physical impairments that come with aging pose a common person-environment disparity. In principle, the gap between demands on the person and the individual's capabilities can be closed, or at least decreased, through suitable control of the environment, the use of techno-logical aids, and human-engineered design of our surroundings, coupled with behav-ioral adjustments. Accordingly, so-called successful aging (D. B. D. Smith, 1990) is often achievable by applying the human factors concept of fitting the environment and tasks to the traits of the person (Fisk, 1999; Fisk et al., 2004).

The role of the aging person in society has been variable at different times and in different regions. At the extremes, an aged person may be considered wise and experienced, a leader or advisor — or useless and expensive, removed from societal life. Look at our own societies for examples: the forced early discharge of airline pilots from their flying duties, the common "going into retirement" of many people in their 60s, and the late-life activities of some politicians.

One might argue that the typical older pilot, for instance, is simply no longer able to fly reliably because of reduced capabilities as a result of the aging process (Morrow & Leirer, 1997). Although one can dispute specifics, in essence this argu-ment sounds reasonable, especially to airline passengers. Other questions go beyond reason but are in the realms of ethics, morals, conscience, and social obligations: What does society owe to an old person? How much care and effort can be expected from family and friends? And what should the recipient do in return? Blaikie (1993); Clements (1993); Fisk and Rogers (1997); Roberto, Allen, and Blieszner (2001); Shephard (1995); and Smolowe (1996), among others, illuminated changes in social context during the 20th century.

From the viewpoint of the ergonomist, designing for the aging has some simi-larities to the enabling of persons with disabilities: As one gets older, many of the faculties diminish that one naturally employed in younger years. In that respect, helping the aging person overcome reduced abilities is comparable to helping individuals with impairments, discussed in Chapter 5.

For example, wheelchair users need specially prepared passageways that are flatter in both slope and smoothness and wider than appropriate for a person who is able to walk; however, paths suitable for wheelchair users are commonly easier to walk along and thus serve older persons as well. The Americans with Disabilities Act (ADA) of 1990 and subsequent documents require many accommodations for persons with impairments, which also make life easier for older persons — and for everybody else. Examples are ample accessible parking spaces, ramps instead of steps, wider entrances to buildings, bars and push levers instead of round knobs on doors, and ticket-vending machines in which displays are easy to see and with controls that one can operate effortlessly (W. L. Wilkoff, P. C. Wilkoff, & Abed, 1998).

The following text summarizes knowledge about changes in our body and mind that we experience as we age. The effects of the aging process have been extensively researched during the last few decades. For more detailed information, one can consult the books by Birren and Schaie (2001), Craik and Salthouse (2000), W. Rogers and Fisk (2000, 2001), S. Smith et al. (2000), and Steenbekkers and Beijsterveldt (1998). Because gerontological research continues to grow strongly, it is prudent to seek the newest publications.

AGING-RELATED CHANGES AND THEIR ERGONOMIC COUNTERACTIONS

To apply ergonomic actions that can be taken to counter age-related declines, we may group the changes into three principal categories: those that determine body size, those that affect primarily the capacity to perform strenuous physical work, and those that affect psychomotor performance primarily related to seeing, hearing, and perceiving other sensory signals.

CHANGES IN ANTHROPOMETRY

Measurement of body dimensions is usually done, in the aging population as in younger adults, with a cross-sectional approach: All available people are measured and then the results combined within certain age brackets (Annis, Case, Clauser, & Bradtmiller, 1991; Annis & McConville, 1996). The resulting averaged data are more useful to describe younger adults because their features do not change very much in the bracketed age span. However, this approach presents a major problem in the description of the aging (as well as of children): Within the age bracket, some persons show little change in anthropometry, whereas, in contrast, dimensions of other people vary rapidly within a few years — for example, in stature because of posture and shrinking of spinal discs; in weight because of changes in nutrition, metabolism, and health; and in musculature and strength because of changes in activity levels.

The age brackets commonly used for surveys of older populations are broad, usually encompassing decades or even longer time spans, as opposed to the common five years in younger cohorts. Thus, aging people with very diverse dimensions and of variously declining health and capabilities are contained in each observation sample.

Cross-sectional reporting of characteristics of groups of aging people with broad age brackets does not provide information about individual differences, but it is done nevertheless because of convenience. Similarly, sorting measured results by chronological age is not a good classification criterion for aging persons (or for children). It would be better to describe traits of the aging by a longitudinal procedure in which changes in sizes and capacities are observed within individuals over many years (Salthouse, 2000), yet no such data taken recently on large populations are available.

Apparently, we mostly have to rely on anthropometric information, reported in the literature on large samples; that is, the result of cross-sectional surveys. Kothiyal and Tettey (2000) compared the anthropometry of older Australians with older Britons. Kelly and Kroemer compiled available data (almost all from the United States) in 1990 (reprinted with some additions in Kroemer, 1997a, 1997b; K. H. E. Kroemer, H. B. Kroemer, & Kroemer-Elbert, 2001) and commented that many samples were small and divided into only a few age ranges. Steenbekkers and Beijsterveldt (1998) collected data on anthropometry, mobility, and strength in the Netherlands; Peebles and Norris (1998, 2000) and researchers at the University of Nottingham (2002) did so in Great Britain; and S. Smith et al. (2000) compiled such data from a dozen countries. As a rule, compilations tend to blur the samples' ethnic origins, regional proveniences, socioeconomic status, health, or other attributes — all codeterminers of anthropometric and biomechanical traits. Nevertheless, collective sets do indicate common trends of aging.

Stoudt (1981) and Barlow and Braid (1990) explained the apparent height loss of about 1 cm in stature per decade that usually comes with aging from the 30s on, as the combined results of

- flattening of the cartilaginous discs between the vertebrae
- general thinning of all weight-carrying joint cartilages
- flattening or thinning of the bony segments of the vertebrae
- enhancement of the S-shape of the spinal column in the side view, particularly in the form of an increased kyphosis in the thoracic area (humpback)
- development of scoliosis, a lateral deviation from the straight line displayed by the spinal column in the frontal view
- bowing of the legs and flattening of the feet.

In terms of body weight, American men as a group usually experience the highest body weight in their 30s and then lose bulk with aging. In a 1990 study, on average, American women were lightest in their 20s but then increased in weight with age, becoming heaviest at about age 60 (Annis et al., 1991). The recent trends in North America and elsewhere toward overweight and obesity probably show up in the aging population as well.

CHANGES IN MUSCULOSKELETAL ATTRIBUTES

In addition to the anthropometric changes that may occur with increasing age, there are numerous alterations in musculoskeletal features. The bones, particularly the long bones, become larger in their outer as well as inner diameters, hence are hollower, and larger pores appear in the bone material. Total bone mass decreases. Together with a change in mineral content, these alterations in bone configuration are the major factors in age-related osteoporosis. As a result, the bones become stiffer and more brittle (Ostlere & Gold, 1991).

Women and sedentary persons are more exposed to this development than men and active people. Proper diet, often with calcium added to offset mineral loss, and physical exercise can partly counteract these developments. Changes in bone structure are associated with an increased likelihood of broken bones as a result of falls or other accidents in which sudden forces and impulses are exerted on the body. Injuries to the pelvic girdle, the hip joint, and the femur are particularly frequent in older women, followed by injuries to the shoulder and arm bones.

The smoothness of the bony surfaces in joints, the thickness of joint linings, the supply of synovial fluids, and the elasticity and resilience of joint capsules and ligaments usually deteriorate with aging. This diminishes mobility in the joints and often leads to painful inflammation, arthritis. Changes in the spinal column often appear in terms of mobility, curvature, and strength (Spence, 1994). Degenerative osteoarthritis with its associated pain is the most-cited cause for limitations in daily physical activities (Kramarow et al., 1999).

Well-used muscles can retain their capabilities into old age, but disuse often accompanied by decreased circulatory supply leads to a degenerative loss of musculature. In such cases, both the diameter and number of muscle fibers are reduced, with slow twitch fibers (Type I) less affected than fast twitch fibers (Type II). This reduces strength per se (A. M. Koontz, Ambrosio, Souza, Buning, Arva, & Cooper, 2004), but the ability to control the remaining muscle mass also worsens because the number of motor units diminishes while the size of the remaining units increases.

Combined with age-related neurological changes and proprioceptor and visual information deficits (discussed later), these events lead to reduced strength capabilities together with slowed and more variable motor coordination. (A nightmarish scenario is that of the proverbial little old man driving his family to church on Sunday morning — when suddenly the power booster of the van's brake system fails.) Persons who remain active can counteract these declines in capabilities to some extent, and exercise has proven to be beneficial into old age (see Ketcham & Stelmach, 2001, for a comprehensive review).

CHANGES IN RESPIRATION AND CIRCULATION

Respiratory capabilities, especially the capacity to move air, diminish with increasing age. The alveoli in the lungs become less able to perform the exchanges of oxygen and carbon dioxide. Furthermore, the intracostal muscles and the chest diaphragm lose some of their abilities to generate breathing space in the chest, hence vital capacity decreases. This is coupled with reduced blood flow and possibly with emphysema, the latter often resulting from smoking.

The elasticity of blood vessels decreases and the resistance to blood passage in vessels may increase due to deposits along their walls. Blood cell production in the bone marrow commonly declines. Blood flow to kidneys, liver, and brain decreases. Thin aging people may have reduced volumes of body fluids.

Heart functions also change. The heart commonly shrinks, and maximal cardiac output is lowered. Elevated heart rate takes longer to return to its resting level, which usually does not change. Neural control of the heart may be impaired (Spence, 1994). Keeping functions up by appropriate physical activities, possibly by employing a supervised program of well-planned regular workouts, is beneficial for all aging persons, even for very old individuals who otherwise would become sedentary and sluggish.

CHANGES IN NERVOUS FUNCTIONS

The ability to cope with events, tasks, and the environment depends on detecting, interpreting, and then responding appropriately to sensory information. In the somesthetic system, the numbers of sensor cells in the skin (dermis and epidermis) decrease, along with collagen and elastic fibers. Receptors such as Meissner's and Pacinian corpuscles become fewer in number. The ability to sense tactile, visual, and acoustic clues from the outside wanes (see the following text). Reduction and alteration of cells, together with diminished arterial and venous flow in the blood vessels, decrease the stimulation and conduction activities in the nervous system. Reduction of sensors and afferent fibers combine with reduced nerve conduction velocity. Neuroanatomical reductions in the brain (see the following subsections) slow the interpretation of arriving signals.

Hence, both sensation (the reception of stimuli at sensors and the resulting neural impulses in the afferent part of the neurons system) and perception (the interpretation of the stimuli) diminish with age, with information processing particularly slowed. This leads to increased variability in reception and integration of, and hence varying responses to, external and internal stimuli (Hayslip & Panek, 1989).

CHANGES IN THE BRAIN

In the brain, the seat of the central nervous system (CNS), many neuroanatomical reductions occur, first observable in healthy persons in their 50s and 60s and becoming often more pronounced in the 70s, and very obvious in many people in their 80s. Certain molecules and cells become increasingly impaired and may even disappear, but these changes are heterogeneous.

Age-related changes have been prominently studied in neurons, of which we have about 100 billion at birth. Recent research findings cast some doubt on the traditional conjecture that the neurons present at birth do not proliferate thereafter but decrease in number as one gets older. However, it is well established that the cell bodies (soma) of neurons and their extensions (axons and dendrites) tend to atrophy. Age-related changes in the neurons of the basal ganglia specifically affect the planning, initiation, timing, and learning of complex motions. Concurrent changes in the cerebellum and motor cortex combine to affect directly the regulation of smooth movements, body posture, and balance (Ketcham & Stelmach, 2001).

The shifts are not limited to neurons: Ganglia cells, which have a supporting role, also change, and the areas between neurons become altered. Extracellular spaces accumulate deposits called *senile plaque*. Fairly little is known about the precise effects of the changes in ganglial cells and associated plaque deposits, yet the dramatically increased deposition of amyloid protein in persons with Alzheimer's disease is alarming. The diverse structural alterations occurring in the aging brain are likely to be associated with the accruing of defects in the DNA, which seems to lower the quality and quantity of critical proteins (particularly enzymes) in cells. This may especially apply to mitochondria, which provide energy to the cells. Other molecular changes have been seen in proteins and enzymes. Chemical changes occur in the membranes of synapses and in the composition of the myelin in axon sheaths (Selkoe, 1992).

Certainly, the brain's aging commonly diminishes abilities for learning, memory, planning, and other intellectual processes and impedes dependent motor functions. Yet, all these diverse anatomical and physiological changes in the mind mean "very little" (Selkoe, 1992) for many older individuals: Among persons between the ages of 65 and 75 years, fewer than 5% show symptoms of dementia, although that number jumps to about 50% in people older than 85 years.

Given enough time and an environment that keeps anxiety at bay, most healthy older people score about as well as young or middle-aged adults on tests of mental performance. The more complex a task is (for example, a multistep mathematical function), the more likely it is that an otherwise healthy elder will perform less well than a young adult. Yet, a message of guarded optimism emerges for many investigations of normal early aging: One may not learn or remember as rapidly as at younger ages, but one can still learn and remember nearly as well (Selkoe, 1992, p. 139).

People employ different coping strategies. One's failing ability to recall names, for example, may be overcome to some extent, and for some time, by developing mnemonic tactics. An aging person may reduce activity boundaries by maintaining only those activities that he or she can handle comfortably and competently. Maintaining physical or mental activities, purposely so for best effect, can counteract reductions in faculty significantly and for long periods.

An extensive literature discusses age-related cognitive functioning with its differential changes in component abilities such as processing speed, working memory, verbal skills, and attention. Handbooks provide insights into and overviews of the current state of knowledge of psychometric abilities, skill maintenance, and proper design of training strategies (see, for example, Birren & Schaie, 2001; Craik & Salthouse, 2000; Fisk et al. 2004; Rogers & Fisk, 2000; S. Smith et al., 2000). The information in such books provides rich information for the ergonomist as a basis for human-engineered design of the aging person's environment.

CHANGES IN THE CAPACITY FOR PHYSICAL WORK

The declines just described in musculoskeletal, respiratory, circulatory, nervous, and mental functions decrease people's capabilities to perform physical work, in terms of both short-term efforts (strength) and lasting efforts (endurance). These declines are often accompanied by difficulties in maintaining postural stability and greater vulnerability to acute or cumulative injuries. Yet, the overall capacity to do physical work is composed of so many components, both physical and attitudinal, that a wide range from strong to weak prevails among aging individuals.

For example, Jackson, Beard, Wier, and Stuteville (1992) showed that in a cross-sectional sample of more than 1,600 men of ages 25 to 70 years, the physical work capacity (measured as maximal oxygen uptake capability — see Chapter 2) was more closely correlated with changes in body composition (mostly body fat content) and exercise habits than with age. Similarly, tolerance of heat (Pandolf, 1991) or cold (K. C. Parsons, 2003; Young, 1991) appears more related to fitness than to age. Thus, the adage "use it or lose it" certainly applies to the aging body: Continuing to do physical work or exercising is of great value in maintaining skills and capabilities.

In addition, ergonomic measures can substantially alleviate the difficulties in performing physical tasks. Among the mitigating measures are the use of well-engineered workplaces, tools, controls, and procedures, and the utilization of powered assistive devices (Amditis, Bekiaris, Braedel, & Knauth, 2003; Cremer, 2001; Haeroe & Vaerinen, 2003; Kroemer & Grandjean, 1997a; K. H. E. Kroemer, H. J. Kroemer, & Kroemer-Elbert, 1997; K. H. E. Kroemer & A. D. Kroemer, 2001a; Kroemer et al., 2001; Marquie & Volkoff, 2001). It is also prudent to reorganize tasks into those that one will continue to do (perhaps in a relaxed manner) and others that will be delegated.

CHANGES IN VISUAL FUNCTIONS

Eye Movements

Ocular motility, the ability to move the eyeball by muscular actions, usually becomes impaired with increasing age. This reduces the ability to perform quick movements, turn the eyes to extreme angles, and focus on a close object. Pursuit movements are impaired in most aging persons, which limits their ability to follow a target smoothly and move the gaze to a target and keep it fixated there. Furthermore, the field of vision becomes smaller at the edges in an older person, particularly in the upward direction (Fozard & Gordon-Salant, 2001; Kline & Scialfa, 1996).

Accommodation Problems

As we grow older, the lens loses water and becomes stiffer. The eye loses some of its accommodation capability as a result, and the near point of accommodation, once usually just 10 cm from the eye, worsens, commonly to about 20 cm at age 40 and to about 100 cm at age 60. This condition is known as *presbyopia*. The problem is that the eye cannot make light rays converge exactly on the retina. If light rays converge in front of the retina, the person is called *near-sighted* (myopic) and finds it hard to focus on far objects but has little trouble seeing close objects. This condition often improves a bit with age because a flattened lens, common with aging, makes the rays from distant objects strike the retina (although not exactly in focus) so that they become recognizable. If the focus point is behind the retina, the person is called *far-sighted* (hyperopic). This condition usually worsens with age, meaning that it becomes more difficult to focus on near objects. Hyperopia is much more common in older people than myopia. Fortunately, use of contact lenses or spectacles can correct both problems fairly easily.

Less Light Passing Through the Pupil

Baggy or droopy eyelids, common with aging, may reduce the amount of light entering the eye. With increasing age, the opening of the pupil becomes smaller, further reducing the amount of light entering the eyeball. This disorder, *senile miosis,* has the most serious effects in dim light. There is a possible benefit from the smaller diameter, however, similar to having a smaller aperture opening in a camera lens: The depth of field may be enhanced, meaning that objects both near and far are in better focus, though they appear dimmer (Kline & Scialfa, 1996).

Cataracts

A cataract may occur at any age but is most common in older people. Water-insoluble dry protein becomes more prevalent, and macromolecules appear in the lens. This decreases transparency and may actually cloud the lens, thus diminishing the amount of light transmitted into the eye. The increased opacity distorts the path of light passing through the lens. Dispersion of light rays at the macromolecules may act like a light veil or film over the eye. (This resembles trying to look through a waterfall, or *cataract,* a term derived from the Greek or Latin.) The effect on vision depends on the macromolecule's size, location, and density. A small cataract in the center of the lens is likely to affect vision far more than even a large macromolecule at the periphery.

Cataracts can cause blurred or double vision, halos and light rays at bright lights, spots in the field of vision, and difficulty in seeing with too little or too much illumination. Replacing the natural lens with a manufactured one can ameliorate the problems.

Yellowing of Lens and Humor

The young eye has a slightly yellow-tinted lens, which makes it act as an ultraviolet filter for the retina. When more fluorescent chromophores of yellow color develop with increasing age, the lens becomes more tinted. A yellowed lens is a stronger light filter, absorbing some of the blue and violet wavelengths. This changes one's perception of colors: White objects appear yellow, blue is hard to detect, and blue and green are difficult to distinguish.

Another problem related to yellowing of the vitreous humor is that the stronger tint causes absorption of more energy from the light passing through; however, increased illuminance of the visual target can help to maintain good visual acuity. Unfortunately, the yellowing problem cannot be corrected with artificial lenses.

Floaters, Veil, and Glare

Liquid and gel portions may clump together in the vitreous humor, causing "floaters" or spots to appear in the field of vision. Pockets of liquid may form. Clumping and liquefaction together can reflect light rays within the vitreous humor, causing the perception of a light veil (similar to mist) in the visual field. If the field contains bright lights, the resulting "veiling glare" can greatly impair one's visual acuity.

Loss of Cones, Rods, or Retinal Pigment

Cones, which are responsible for color vision and general visual acuity, operate only under high-illumination (photopic or bright) conditions. Cones are most densely concentrated in and around the center of the retina, the fovea. There is little age-related loss in foveal cones, but their number at the periphery can diminish.

Rods provide white and gray vision even in low-light (scotopic or dim) conditions. Rods are located around the fovea, extending to the peripheral areas of the retina. Their number can decline dramatically with aging, especially near the center. Retinal pigment that provides for the metabolism of the photoreceptors is reduced as well.

Combined, these degenerations reduce the clarity of vision (acuity) in general and of peripheral acuity in particular. The age-related changes raise the threshold of illumination needed for vision and make the discrimination of colors difficult, especially between blues and greens but less so between yellows and reds.

Self-Reported Vision Problems

Common self-reported vision problems (Kline & Scialfa, 1996) among the aging include decrements in acuity, near and distant vision, visual search, contrast sensitivity, seeing in shaded areas, seeing (and hence driving) in twilight, tolerance of glare, seeing when entering or leaving dark areas, spatial vision, and color vision. Objectively measurable dysfunctions, just discussed, can explain some of these difficulties. Other visual problems, however, appear to result from faults in the visual apparatus that interact with associated nervous and brain processes. Such decrements may be too small or otherwise difficult to detect in laboratory tests.

Recommended Checkups

Eye examinations can detect vision problems at their early stages and trigger behavioral, ergonomic, and medical countermeasures that help to prevent damage in many cases. In addition to the age-related developments listed already, especially dangerous are sudden jolts or vibration that may detach the posterior vitreous from the retina. Another severe vision impairment may stem from natural fluids that may not drain well but collect inside the eyeball. The ensuing pressure (characteristic of glaucoma) may eventually destroy fibers of the optic nerve. Glaucoma is a leading cause of blindness in older people. Regular eye examinations, recommended especially after the age of 40 years, help to detect the onset of internal pressure and to initiate treatment that prevents damage.

CHANGES IN HEARING

The ability to hear decreases dramatically in the course of one's life, first in the high frequencies between 20 and 10 kHz, then even lower to about 8,000 Hz, an impairment commonly known as *presbycusis* (age-related hearing loss). Yet, difficulties in hearing may extend into even lower frequencies, and they are often coupled with noise-induced hearing loss, already discussed in some detail in Chapters 1 and 2.

Although it is estimated that 70% of all individuals over 50 years of age have some kind of hearing loss, the changes are different for each individual. Decrements attributable to the environment overlap with, and in some cases mask, age-related decreases. Typically, in populations that are not exposed to job- or civilization-related noises, the hearing sensitivity to higher frequencies is less reduced than in people from developed countries.

The changes start at the pinna, the outer ear, which becomes hard and inflexible and may change in size and shape. Wax buildup is frequent in the ear canal. Often, the Eustachian tube becomes obstructed; this can lead to an accumulation of fluid in the middle ear, which normally contains only air. There may also be arthrotic changes in the joints of the bones (anvil, hammer, and stirrup) of the middle ear, which, however, do not usually impair sound transmission to the oval window of the inner ear. There, in the cochlea, atrophy and degeneration of hair cells in the basilar membrane commonly occur. Deficiencies in the bioelectric and biomechanical properties of the inner ear fluid and mechanical degeneration of the cochlear partition also appear, often together with a loss of auditory neurons. Such degeneration causes either frequency-specific or more general deficiencies in hearing capabilities.

Loss of hearing ability in the higher frequencies of the speech range especially reduces the understanding of consonants that have such high-frequency components. This explains why older persons often cannot discriminate between phonetically similar words, which may make it difficult to follow conversations in noisy environments. Severe hearing disorders may lead to speech disorders, which, to some extent, may be psychologically founded. For example, if people have to speak loudly for an old person to understand, they may not want to interact with that person to avoid embarrassment. There may be hesitancy to speak if it is uncertain what level of loudness is required and that one understands what is said.

CHANGES IN TASTE AND SMELL

The number of taste buds and the production of saliva in the mouth often reduce with advancing age. Fissuring of the tongue may occur. The sense of smell often diminishes, although this finding is not uniform across studies (Belsky, 1990; Birren & Schaie, 2001; Hayslip & Panek, 1989). Altogether, a precise understanding of changes in gustation and olfaction is hampered by the lack of detailed knowledge about these senses, discussed in Chapter 2. Although reductions in sensitivity to smells and tastes may simply make the environment less stimulating and hence more boring, adding more tasty ingredients to food and drink, and more intensive smells to the environment, may make up for the loss in sensitivity.

CHANGES IN SOMESTHETIC SENSITIVITY

The somesthetic (kinesthetic) senses include those related to touch, pain, vibration, temperature, and motion. In spite of decades of research, there is deplorably little reliable and quantitative information about the changes with aging (Boff, Kaufman,

& Thomas, 1986), in no small measure because of the paucity of up-to-date research (see Chapter 2). Apparently, absolute thresholds increase, which may be associated with a loss of touch receptors, but that phenomenon and its explanation still require more investigation. The same is true for pain sensitivity, which appears much reduced as one gets older.

Vibratory sensitivity, particularly in the lower extremities, commonly decreases. This effect is used in the diagnosis of disorders of the nervous system. Decreased information from kinesthetic receptors or reduced use of that information in the CNS may contribute to the higher incidence of falls: With increasing age, one seems to be less able to perceive that one is being moved (such as in an automobile or airplane) or that one is actively moving body parts. Again, surprisingly little systematic information is known. Possible explanations are reductions occurring in the number of Meissner's corpuscles and other skin receptor organs and the number of myelinated fibers in the peripheral nervous system (PNS) and in blood supply (Committee on an Aging Society, 1988; Hayslip & Panek, 1989). Dietary deficiencies might also play a role.

Most older persons do not tolerate a cold environment well. Their "temperature behavior" may be associated with a decline in the ability to regulate the body's temperature. Temperature sensitivity also seems to decline, but this may be partly offset by the often-observed desire of older persons to be in warmer temperatures, outside or within a building. In a colder climate, old persons may be in severe danger if they do not recognize their falling body temperature. Yet, although a review of the literature on tolerance to cold and heat suggests declines with age, these effects may be more closely correlated with health, fitness, and acclimatization than with aging per se (Pandolf, 1991; K.C. Parsons, 2003).

CHANGES IN SENSORY AND PSYCHOMOTOR PERFORMANCE

Although many skills can be retained by practice and with good health, others inevitably deteriorate with age due to diminished circulatory and metabolic capabilities because of physical and chemical changes in the body's musculoskeletal components, in the CNS and the PNS, as discussed earlier in this chapter. Even in healthy elderly and old persons, reductions are common in activities that require exertion of energies or forces controlled through perceptual information, particularly of the visual and vestibular senses. (S. Smith et al., 2000; Steenbekkers & Beijsterveldt, 1998; Vanderheiden, 1997; and Vercruyssen, 1997 provided overviews.)

Manipulation is a skill that particularly incorporates mobility and sensory control. In spite of the fact that manipulation skills are important for many tasks on the job, at home, or during leisure, unified and standardized measuring techniques are not available, although progress is being made (Scott & Marcus, 1991). A comparison of information on hand manipulation capabilities of adults in general (Daams, 2001; Weimer, 1995) and of aging persons (Imrhan, 2001; Peebles & Norris, 2000; S. Smith et al., 2000) supports the experience that severe declines in manipulation capabilities are common, often associated with arthritic joint diseases.

With aging, time needed to react and respond to stimuli generally increases (see Chapter 2 for definitions and measures). This may be partly explained by deficiencies in the sensory (afferent) peripheral parts of the nervous system, delays in afferent provision of information to the CNS, and reduced efficiency in the motor (efferent) part of the PNS. Yet, successful performance involves not only motor behavior but also perception, sensation, attention, short-term memory, decision making, intelligence, and personality. Thus, although poor motor performance may be attributed overtly to some of the physiological factors, impeded processes of the mind may be strongly implicated as well.

The changes in seeing, hearing, tasting, smelling, and somesthetic senses that occur with aging, as outlined earlier, though individually variable (W. A. Rogers, 1997b), in general are apparently unavoidable and irrevocable: There is no evidence in the literature that these sensory abilities can be retained by use or training. Fortunately, the commonly prescribed seeing and hearing aids often effectively counteract some losses. Furthermore, purposeful human engineering can compensate for many of the reduced sensory abilities. Examples mentioned in Chapter 3 are distal signals in the form of lights, colors, and sounds that can be magnified, modified, or otherwise transformed so that aging human eyes and ears can perceive these inputs.

The environment can be controlled with respect to temperature and humidity and ergonomically arranged to provide suitable lighting; for example, in buildings, offices, automobiles, and at computer workstations. Sound signals can be selected in their intensity and frequency components to be clearly perceived, and noise can be suppressed (for more information, see the following sections in this chapter and Kroemer & Grandjean, 1997; Kroemer et al., 1997; K. H. E. Kroemer & A. D. Kroemer, 2001a; Kroemer et al., 2001; W. A. Rogers, 1997a; Rogers & Fisk, 2000).

DESIGNING FOR THE OLDER PERSON

The U.S. Age Discrimination legislation considers someone of 40 or more years of age an "older" person. Although no one given year establishes the beginning of aging, there is little doubt that certain work tasks become more difficult as one's age exceeds the middle 40s. Tasks that require high visual acuity and close focusing become harder, as do tasks that require high mobility, particularly of the trunk and back, and so do strenuous physical exertions such as moving heavy loads. On the other hand, scientific findings (collected, for example by Birren & Schaie, 2001; Craik & Salthouse, 2000) and even more anecdotal evidence indicate that many (or at least some) older workers can perform tasks demanding patience and experience-gained skills better than many young persons.

As long as age-generated difficulties remain fairly small, thoughtful human engineering can overcome or reduce them by proper workplace design and tool selection, suitable arrangement of work procedures and tasks, and provision of working aids such as power-assisted tools or magnifying lenses. (Several such ergonomic means to mitigate age-related losses were mentioned earlier.) Organizational and managerial measures can be of great benefit as well; for example, considerate job and shift assignment, and provision of work breaks (Fisk & Rogers, 1997; Haermae, 1996; Haex & van Houte, 2001a, 2001b; Hammer & Price, 2001; Karazman, Kloimueller, Gaertner, Geissler, Hoerwein, & Morawetz, 1998, Kumashiro, 1995;

K. H. E. Kroemer & A. D. Kroemer, 2001a; Kroemer et al., 2001; Small, 1997; Shephard, 1995). Suitable software may facilitate work with computers that requires significant involvement of short-term memory and that demands manipulative skills (Czaja, 1997, 1998, 2001a; Czaja & Lee, 2001; Fisk et al., 2004; Morrell & Echt, 1997).

The ergonomist can facilitate manual work by providing proper work height, supports for arms and elbows, and provision of special hand tools, power-assisted if needed. Postural aids include the just-mentioned proper work height coupled with provision of a seat with appropriate dimensions and easy adjustments. For the chair, in essence the same design recommendations apply whether it is used in the shop or office, but the shop seat must be especially sturdy and protected against soiling (K. H. E. Kroemer & A. D. Kroemer, 2001a; Kroemer et al., 2001).

Ergonomic interventions can ameliorate many sensory problems as well. For example, using corrective eye lenses, providing intense and well-directed illumination, and avoiding direct or indirect glare (more on glare later) may offset vision deficiencies.

Keeping background noise at a minimum improves the auditory environment so that speech and important auditory signals have sufficient "penetration" (Huey, Buckley, & Lerner, 1994; Lerner, & Huey, 1991). This is especially important for emergency alarms, which should have appropriate sound intensities and frequencies, avoiding frequencies higher than 4,000 Hz.

Hearing aids that amplify sounds for which hearing deficiencies exist can dramatically improve a person's ability to perceive acoustic signals and words. Providing such a hearing aid, however, is difficult if the deficient auditory ranges are not precisely known: Quite a few persons are hesitant to take a hearing test. One problem is that the amplification of sounds might also amplify unwanted background noise. Furthermore, hearing aids often look clumsy; are difficult to insert, adjust, and maintain; may malfunction or stop working just when needed; often whistle annoyingly; and can be bothersome to use with a telephone (Barnes & Wells, 1994). Technical development may improve these devices in the future: Built-in active noise suppression by interference is now becoming practical at least for frequencies below 1,000 Hz (Berger, Berger, Royster, Royster, Driscoll, & Layne, 2003). Former President Clinton's well-publicized wearing of an in-ear hearing aid was a step toward reducing the general aversion to admitting to a hearing deficiency and employing a technical device to improve the condition — similar to wearing glasses to enhance vision.

Simple ergonomic measures can improve the clarity of the auditory message: provision of sound signals that are easily distinguishable by sufficient intensity and suitable frequency and avoidance of masking background sounds. Another solution is to provide information through other sensory channels; for instance, using flashing lights together with a wailing siren or employing vibration of a body-worn phone to indicate an arriving call.

Selection of proper words, fonts, and contrast for signs, labels, and instructions on ticket-vending and teller machines and on medicine bottles are examples of good ergonomic practices that make usage easy. Such designs help older persons, nonnative English speakers — and everyone else. In essence, use of proper ergonomic measures, carefully selected and applied to help older persons, is just good human factors engineering.

DESIGNING FOR THE AGING EYE

Aging is associated with the developing deficits in basic visual functions listed earlier: light sensitivity, near vision, depth perception, and dynamic vision all deteriorate together with perception and processing. Individuals develop visual impairments of different types and magnitudes at varying ages, but common experiences are difficulties in seeing in dim light, reading small print, distinguishing similar colors, and coping with glare. There are some simple solutions for most of these problems: for example, providing proper corrective lenses, higher intensity of lighting, and increased color contrast; use of large characters with high contrast against the background; repositioning a computer screen; or shielding bright lights in the field of view. These solutions are discussed in detail in ergonomics books by Schieber (1994), Konz and Johnson (2000), K. H. E. Kroemer and A. D. Kroemer (2001a), and Kroemer et al. (2001), and in other publications on human factors engineering listed in Chapter 1.

The aging human eye loses precision, to borrow a term used with cameras. Changes occur in the structures that bend, guide, and transform light. This reduces the amount of light reaching the retina and blurs the image projected onto it. The aging eye has trouble in focusing on near objects, particularly if they are elevated or move fast. Visual acuity, contrast sensitivity, and related spatial vision capabilities, especially with a busy background and dim lighting, decline with age. Shrinking pupils, yellowing lenses, cataracts, and droopy eyelids, common in older persons, mean that less light strikes their retinas compared with younger eyes. Therefore, many older people must have increased illumination on visual objects for sufficient visual acuity — but that means an increased chance for glare. Glare is in essence the effect of high-intensity light, often in sharp contrast with its darker surroundings, that overpowers the vision system and makes seeing difficult. Direct glare meets the eye when a light source such as the sun or a lamp in the office shines directly into one's eye. Indirect glare is light reflected from a shiny surface, such as the sun or a lamp mirrored on the surface of the computer display.

Different optical eye lenses (spectacles or contact lenses) may have to be used at video display workstations and for reading and writing tasks, particularly if the display is farther away than easy reading distance, which is normally at about 40 cm. Because low-contrast images are hard to see, aging workers may have difficulty reading displays that have a dark background. Recognizing details from a cluttered background becomes hard. Picking out details from similar objects (or detecting individual faces in a crowd) in dim light is nearly impossible. Even reading a text in large print may be laborious if only a little white space exists between the black letters. If letters or graphs are in color, they should be in bright reds and yellows that are easier to distinguish than blues and greens. Computer displays should be located distinctly below eye height to facilitate accommodation and convergence of the aging eye, as discussed in the following text on computer system design.

Extra lighting to increase illumination can improve the visual ability of aging persons. However, because glare is a problem for many, task lamps and other bright light sources must be placed carefully. Aging people often find it difficult to adapt to sudden changes in lighting, particularly from bright to dim conditions.

DESIGNING FOR THE AGING DRIVER AND PASSENGER

Depending on whose statistics one reads, aging drivers may have fewer or more traffic accidents than their younger colleagues (Laflamme & Menckel, 1995), but the pattern of accidents that accompanies aging indicates missed or misinterpreted perceptual cues, slow or false reactions to cues, and wrong motor actions such as accelerating instead of braking. There is a vast literature on the many behavioral traits of the aged driver, such as compiled by Karwowski (2001), Karwowski and Marras (1999), W. A. Rogers (1997b), and Salvendy (1997). In 1990, Koppa provided an extensive overview of adaptive equipment in automobiles for use by aging persons and by those with disabilities. Given this abundance of literature, here I will indicate only general ergonomics principles and some specific human engineering applications that, as usual, benefit both the aging and the younger driver and passenger.

Many aging people drive automobiles, although they may avoid driving at night, on unfamiliar roadways, in congested areas, and during busy times. Vision is especially a problem in night driving (Noy, 2001; Olson, 2001; Wood, 2002), made worse by a clutter of lights and colors, present in many commercial streets where high general illumination and especially red and green advertising lights compete with traffic lights (Ho, Scalfia, Caird, & Graw, 2001).

Regarding public transportation (planes, trains, subways, buses, trams, lifts, elevators, moving walkways, etc.), aging persons report — as do many younger people — problems with cues and signs that should provide information about use, direction, and location, such as

- How do I buy a ticket?
- Where does the bus go that I see coming?
- Where does the tram stop?
- Where is the exit to Brown Street?
- Where are we?

Many of these problems can be overcome by applying common human engineering principles, such as using signs with good lettering (contrast, size, and symbols) and proper illumination, and avoiding glare. Auditory announcements must be timely and understandable. Redundant information is commonly helpful — for example, showing the next floor number in an elevator in large numerals and announcing it early over a public address system.

Another set of major problems for older persons relates to ingress into and egress from transportation vehicles such as trains, trams, buses, and passenger cars. Entrance and exit passages and steps are often difficult to maneuver and may require forceful and complex stepping, typically so in trams, trains, buses, or moving walkways. Uneven floors and damaged or misplaced floor coverings often pose problems in vehicles or hallways. Handholds used to be a problem in buses and trams, but most now provide a variety of columns, horizontal bars, hooks, and handgrips.

In one survey, 12% of older men and 18% of older women reported not being able to walk a quarter mile; yet, among persons over 70, only 17% used a cane and

10% used a walker (Kramarow et al., 1999), and some needed wheelchairs. A cane usually does not pose a problem in public transportation (in fact, it may alert other passengers to be considerate and helpful), but walkers and particularly wheelchairs are often not easily accommodated and may be a serious deterrent to the use of public transportation.

As mentioned in Chapter 5 with reference to persons with disabilities, the use of ergonomics knowledge in the design of assistive devices ranges from walking aids to special automobiles, from wrist splints to artificial limbs — see, for example, publications by Fisk and Rogers (1997), Karwowski and Marras (1999), Kroemer et al. (2001), or Salvendy (1997). Often, technology used to assist persons with disabilities also helps people who have lost their former capabilities to the ravages of aging. Conversely, design to suit the aging also is useful for persons with disabilities, such as the design of the habitat discussed later in this chapter.

Many everyday consumer products could be better designed than they are; for example, household utensils, knife handles, remote controls, child-proof medicine bottles, door openers, vacuum cleaners, stoves, razors, faucets, audiovisual equipment, or computer keyboards. Ergonomic design would make these items easy to use for any person, young or old, whether or not impaired by age or injury. Often, relatively small changes in size, shape, control, direction, and effort of operation (Daams, 2001; Freivalds, 1999) can render a product usable and safe (August & Weiss, 1992; Fisk, 1999; Fisk et al., 2004; Nayak, 1995; Pirkl, 1994).

DESIGNING COMPUTER SYSTEMS FOR THE AGING

Computers have opened the door for many persons who would otherwise feel shut in; electronics allow them to easily communicate with other persons, to shop and to bank. However, computer technology can be a challenge as well — often bothersome, sometimes welcome — to many older persons for building and maintaining their skills, and keeping them in contact with others and the life around them while enhancing their self-confidence and independence. A concern of the 1990s was that many elderly and old persons appeared hesitant to become acquainted with the then-new computer technology, but this problem largely disappears with the passage of time as computer-cognizant persons enter the aging population and as designers consider the special needs and preferences of older users (Goebel, Fietz, & Friedorf, 2003).

As mentioned in Chapter 5, computers are

- relatively easy to use as an input or output device
- capable of processing complex information rapidly
- able to control other equipment.

These traits make them particularly helpful for many aging persons. The customary QWERTY keyboard is ergonomically faulty (Kroemer, 2001), but to use it, one does not have to be able to write by hand. However, it requires extremely repetitive motions of the hand and the digits, which may overstrain the aging musculoskeletal system (K. H. E. Kroemer & A. D. Kroemer, 2001a; National

Research Council, 1999; National Research Council and Institute of Medicine, 2001). A theoretically better solution is speech recognition software, some of which, unfortunately, is still not easy to use because of imperfections in terms of correct interpretation of spoken inputs and spelling (Chapter 5 contains more on this topic).

Developments in computer technology have facilitated communication among people, which is especially important if it is not feasible to visit each other or travel to distant destinations. They have also opened up the Internet and thereby made available an abundance of information to persons who otherwise would be isolated because of disability due to age, illness, or injury. However, many potential users encounter problems in the use of the equipment because some system designers still seem not to sufficiently consider persons with age-related impairments as active users: Ergonomic deficiencies in keyboards and other input devices, as well as output devices, or inadequacies in manuals and instructions are particularly annoying and hinder efficient operation (Jamieson & Rogers, 2000; Kroemer, 2001; K. H. E. Kroemer & A. D. Kroemer, 2001a; Mead, Batsakes, Fisk, & Mykitsyshyn, 1999; Mead, Sit, Rogers, Jamieson, & Rousseau, 2000; W. A. Rogers, 1997a, 1997b; W. A. Rogers & Fisk, 2000). Cleverly designed software can facilitate work with computers by reducing the demands on short-term memory and on manipulative skills. Proper computer technology can effectively link clients, their caregivers, and medical staff (Czaja, 2001a).

Designing the computer workstation for the older user is, in principle, the same as setting it up for any adult computer operator (such as described in some detail by K. H. E. Kroemer & A. D. Kroemer, 2001a, 2001b; and Kroemer et al., 2001). However, the ergonomist must use special care to accommodate any particular needs related to posture, mobility, and vision of the mature person. For posture, a carefully selected supportive chair is necessary. Which specific chair is comfortable depends on the individual user, but as a rule, a stiff seat enforcing an upright posture is not generally suitable. Instead, most users want to select from modern chairs that support a variety of body movements between tilting forward and leaning back (HFES, 2002). The use of a wheelchair necessitates special requirements regarding the spatial arrangements of the workplace, especially for moving in and out and for proving space for the legs while permitting easy access to the work equipment. If hand mobility is restricted, then special input devices are required, similar to those discussed in Chapter 5 for use by persons with impairments. Speech recognition software may help those with reduced joint mobility in hand, arm, and shoulder, often caused by age-related arthritis.

For easy vision, the computer display should be carefully placed at the user's individually preferred seeing distance (with reading lenses worn). The binocular accommodation and vergence positions to which the eyes return at rest vary from person to person but are specific for each individual. Thus, each computer user should be encouraged to move the screen to her or his most comfortable distance. A point of particular concern is to ensure correct height placement of the monitor: not above or at eye height, as occasionally but wrongly recommended (see the discussion in Chapter 5) but somewhat below eye height. The reason for avoiding high screen positions is that many older persons, sitting or standing, find it difficult to raise their eyesight (Kroemer, 1993) especially while trying to maintain a short vergence distance, which is needed to distinguish fine details. Therefore, it is difficult for most adults, and especially for aging persons, to focus on near objects such as script on a monitor when they appear above eye height. Generally, it feels more comfortable to lean back in a chair while tilting the head down to look at a close object (similar to reading a book) than to sit upright: Bending the neck forward and tilting the eyes downward makes it easiest to focus on the display or print because it reduces the natural vergence distance. Usually, the best and simplest solution is to place the display as low as possible, directly behind the keyboard, as with a laptop computer.

Shrinking pupils, yellowing lenses, cataracts, and droopy eyelids, frequent in older persons, mean that less light strikes their retinas compared with younger eyes. Therefore, many older people must have increased illumination on visual objects for sufficient visual acuity. Task lights shining on printed texts to be read are often useful, but resulting direct or reflected glare striking the eye should be avoided.

DESIGNING INTERNET HEALTH INFORMATION

Seniors increasingly use Web sites to obtain health-related information and interact with medical and service providers. Among the principal areas of interest are definitions of diseases, medical terms, drug uses, drug interactions and side effects, herbal remedies, health guides, caregivers, health care plans and benefits, health care organizations, research, insurance, supplies, and emergency actions (Kwahk, Smith-Jackson, & Williges, 2001). Such a determination of areas of needs and interests is the first step in the development of telesupport systems (discussed in more detail in Chapter 5). This information provides the basis for the ergonomic design of the content of an electronic information system.

The next step is designing the interface so that it meets aging users' physical and cognitive capabilities, their preferences, and usage patterns. For example, seniors may have motor control problems that make mouse control difficult; they may need large, bold-font text that is easy to read on a high-contrast background; and they may not want flashing and other dynamic display features that distract and confuse.

These initial tasks — first the determination of needs and then the designs of content and interface — should be done in close cooperation with the future users. The final task is usability evaluation, which requires the participation of the seniors in carefully planned yet realistic tests held in the users' daily environment, not in a laboratory, that yield real-world information. Smith-Jackson and Williges (2001) thoughtfully discuss these procedures in detail.

DESIGNING LIVING QUARTERS FOR THE AGING

A 50-year-old who purchases a durable product such as a home will be a different individual 10, 20, even 40 years later when still using it. That person may even then be able to live alone in the home, perhaps with some outside support, or he or she might require frequent assistance. (Of individuals 65 and older, one in three lives alone, according to the U.S. Census Bureau in 2003.) The ability to live independently, without need of help, or, conversely, the necessity to live in a care environment — to

DISORDERS

PROBLEMS/ MANIFESTATIONS	Senescence	Arteriosclerosis	Hypertension	Parkinson's disease	Peripheral neuropathy	Drowsiness	Cataracts/glaucoma	Arthritis	Paget's disease	Osteoporosis	Low back pain	Bronchitis/emphysema	Pneumonia	Diabetes	Senile dementia
General debility	☆		☆	☆	☆			☆	☆	☆	☆	☆	☆	☆	
Mobility	☆	☆		☆	☆		☆	☆	☆		☆				
Posture	☆			☆			☆	☆	☆	☆	☆				
Pain		☆		☆	☆			☆	☆		☆				
Incoordination		☆		☆	☆		☆	☆							
Reduced sensory input	☆	☆			☆	☆	☆								
Loss of balance	☆	☆		☆			☆								
Reduced joint mobility								☆	☆	☆	☆				
Weakness in muscles	☆			☆	☆			☆							
Auditory disorders	☆					☆		☆						☆	☆
Locating body in space		☆					☆	☆							
Shortness of breath		☆		☆								☆	☆		
Deformity								☆	☆	☆					
Memory impairment		☆													☆
Visual problems	☆	☆					☆							☆	
Disorientation		☆		☆											☆
Loss of sensation	☆				☆	☆									
Cognition disturbance		☆		☆											
Incontinence		☆													☆
Speech disorders				☆		☆	☆								
Touch disabilities	☆				☆			☆						☆	

FIGURE 6.1 Problems arising from common age-related disorders. (From *Ergonomics — How to Design for Ease and Efficiency* (2nd ed.), by K. H. E. Kroemer, H. B. Kroemer, and K. E. Kroemer-Elbert, 2001, Upper Saddle River, NJ: Prentice Hall. Copyright 2001 by Prentice Hall. Reprinted with permission of Pearson Education, Upper Saddle River, NJ.)

state the extremes — depends on complex interactive individual abilities or disabilities. Figure 6.1 presents an overview of disorders common among the aged and their effects on everyday activities. Many of the resulting problems experienced in daily life can be alleviated, at least to some degree, by proper ergonomic measures.

As discussed in Chapter 2, functional ability (or its opposite, dependency) is commonly assessed in clusters of activities: the instrumental activities of daily living (IADL) and the less specific activities of daily living (ADL), both listed in Table 6.1.

**TABLE 6.1
Measures of a Person's Ability to Live
Without Assistance**

Activities of Daily Living (ADL)
- Dressing
- Toileting
- Bathing
- Transfers between bed and chair
- Indoor and outdoor mobility

Instrumental Activities of Daily Living (IADL)
- Light housework
- Meal preparation
- Doing laundry
- Managing money
- Taking medication
- Making phone calls
- Shopping

Numerous surveys on older persons have used these two listings. The results have served to classify people into groups that require help and specifically designed environments (see, for example, Committee on an Aging Society, 1988; Lawton, 1990; D. B. D. Smith, 1990).

Although ADL and IADL appear practical and self-explanatory, they unfortunately lack specificity and objectivity and are difficult to scale. One attempt to improve the rather gross ADL was to subdivide them into more specific tasks such as lifting or lowering, pushing or pulling, bending or stooping, and reaching (Clark, Czaja, & Weber, 1990; Pennathur, Sivasubramaniam, & Contreras, 2003). Further work in this direction (as done, for example, by W. A. Rogers, Meyer, Walker, & Fisk, 1998) could result in a list of basic demands and activities somewhat similar to the motion elements used in industrial engineering for method studies (see, for instance, Niebel & Freivalds, 1999). Chapter 2 addresses the needs for and possibilities of assessments that are specific, quantitative, relevant, and practical.

A HOME SUITABLE FOR OLDER PERSONS

Living in one's own home has the major advantage because one is in a familiar setting with all its physical and emotional implications. These include feeling at home, being comfortable, enjoying privacy, and having the satisfaction of self-sufficiency and independence. Unless by happenstance or foresight private homes are designed to be ergonomic, they usually need some adjustments to allow the aging resident to perform all necessary activities even while sensory, motor, and decision-making capabilities are diminishing. In addition to passage areas such as stairs and hallways, there are several rooms of particular concern: kitchen, bathroom, and bedroom. The activities that are most important to older adults take place in these rooms, and most accidents happen here.

In the United States, perhaps the best-known house expressly built for use by an older resident with disabilities is the Top Cottage in Hyde Park, New York, which President Franklin Delano Roosevelt designed to accommodate himself in his wheelchair. Fortunately, not all old persons need wheelchairs continuously, but they may have to use them at times. In addition, designing a house for a wheelchair user will certainly make it convenient for any person who is not as agile and powerful as in one's youth. (Chapter 5 contains more information about wheelchair ergonomics.)

A large number of excellent books on age-friendly building and modifications of habitats are available: The American Association of Retired Persons (AARP) provides free brochures and listings (http://www.aarp.org). For private homes, a variety of publications contain valuable ergonomic design recommendations. Among the commercially available books, those by T. Koontz and Dagwell (1994), Peloquin (1994), and Wylde, Baron-Robbins, and Clark (1995) typify the thinking of modern Western architects. In other civilizations and parts of the world, however, different customs and conditions exist for which at present little ergonomic information appears to be available (Cai & You, 1998; Ogawa & Arai, 1995; Pinto, De Medici, Van Sant, Bianchi, Zlotniki, & Napoli, 2000).

Passages within the dwelling and to and from it must be safe and easy to use even for a frail person. The floor surface should be flat, without barriers such as stairs or thresholds, and best not sloped; doors and passageways should be wide enough to allow a wheelchair to pass and turn; and flooring should provide enough friction even when wet. Passages must be well illuminated, as should be all other rooms of the dwelling.

Flights of stairs, steps, and thresholds can make moving about difficult, often hindering persons with mobility restrictions from using all available space, and wheelchair users may not be able to roll up or down at all. If the residence spans two floors, lifts and elevators make life easier. Elevators connecting the ground floor with the next one are fairly easy and relatively inexpensive to set up, especially when the dwelling is planned to accommodate them and installation done early during construction or remodeling. Even low sills at doorways or shower stalls can be a nuisance and cause stumbling, as do the rims of carpets and loose rugs.

Doors and windows must be easy to open and close, even when an additional screen or storm door is present. They require clear space in front to provide access. Controls must be handy and require little strength to operate yet provide security. Push bars and lever handles are easier to operate than round knobs.

Electrical switches and all other controls as well as electrical outlets should be located at about hip height so that persons both standing and sitting in a wheelchair can reach them naturally. They should be easy to operate, at best by a simple push, and not require fine fingering. Automatic lights are advisable in bathrooms, bedrooms, and passageways.

Cabinets and other storage facilities should be within easily reachable distance and height, not requiring excessive body stretching, bending, or twisting. The contents should be within sight, meaning that shelves should be at or below eye height and not so deep that items located in front obscure those at the back.

Climate control can be of great importance especially when the outside conditions are extreme in terms of temperature and humidity. Automatic settings are preferred because they require no judgment, decision, or action by the person. Many people prefer a floor heating system with its uniform warmth to the often drafty and loud forced-air arrangement common in North America.

Kitchen

The kitchen is one of the most frequently used and important rooms for most people. This is not only the location to prepare, serve, and store food, but it is often a gathering room, a social place, and a phone and message center. Lillian Gilbreth completed the first scientific study of kitchens in the 1920s (reported, for example, by Niebel & Freivalds, 1999). Her classic study put special emphasis on the best flow of activities in the kitchen, which she determined with the time-and-motion study methods that she had pioneered with her husband.

Seven human factors principles derived from time-and-motion studies, augmented by newer ergonomic findings, apply to kitchen design:

1. A small work triangle is most efficient and underlies the design concept. Its corners are the three primary activity areas: storing (refrigerator, cabinets, etc.), cooking (gas or electric burners, conventional or microwave oven, and coffee/tea maker), and the cleaning area (sink, dishwasher, etc.).
2. Kitchen design and components should facilitate the workflow for food handling at these activity areas, and among them, and for serving the prepared food.
3. Items should be stored at the point of first use. Rollout shelves are advantageous at low height; generally, shallow shelves are beneficial. Items on storage shelves and in cabinets should be easy to reach and see, not requiring excessive body bending, stretching, or twisting.
4. The workspace for the hands, such as at counters and sinks, should be at about elbow height or slightly below. This facilitates manipulation and visual control. Counter and sink may be put lower than usual for wheelchair-bound persons.
5. Stove, oven, refrigerator, and dishwasher openings should be at "no-bend" heights.
6. The use of wheelchair and walking aids often strongly influence kitchen design.
7. Traffic of others should not cut through the patterns of work triangle and workflow.

Bathroom

Bathrooms are areas of major ergonomic concern because they are essential for healthy living. Basic equipment includes bathtub, shower, toilet, and washbasin. Furthermore, bathrooms usually contain storage facilities for toiletries, towels, and so forth. Unfortunately, many traditional bathroom designs in the United States are difficult to use for aged people and persons with disabilities (Malassigne & Amerson,

1992; Mullick, 1997). Major problems are narrow doors and space so tight that many older users who need canes, walkers, and especially wheelchairs find it difficult to move about.

Bathtub and shower, the two common areas for cleansing the whole body, are sites of numerous accidents. Their major danger stems from the slipperiness of bare feet on wet surfaces. The more dangerous of the two is the bathtub because of its commonly slanted surfaces combined with high sides above which one has to step, a procedure not easy for most people and particularly difficult for older people who have balance and mobility deficiencies.

In his classic studies, Kira (1976) described several techniques to get in and out of a tub. They involve shifts in body weight among legs and arms and buttocks, with much less potential for loss of balance, slips, and ensuing falls. For the resting position in the tub, the angle of the backrest and its slipperiness are the most critical design concerns. Proper handrails and grab bars within easy reach, both for sitting and getting in and out, are of great importance. A shower stall is easier to use because its lower enclosure rim makes it easier to enter and exit. The lip of the shower enclosure should not be higher than 2 or 3 cm, but a rimless design is the best for a wheelchair to roll in and out of a shower stall.

Using the control handles for hot and cold water is often difficult for aged persons, particularly when they have to reach across a tub or shower basin to access them. This can become a real problem for older persons when they are not at their familiar home and have to cope with different handle designs, movement directions, and varying resistances. In some set-ups, the controls for hot and cold water turn in the same direction; other designs use faucets that turn in opposite directions. Better human-engineered design solutions than so far employed, and their standardization, would be advantageous, such as in the mode and direction of control movement to regulate water temperature. To prevent scalding, putting the control handles within easy reach and thermostatically adjusting the water temperature are good ergonomic solutions, helpful to all users. Freivalds (1999) provided extensive information on proper control design.

The washbasin may be difficult to use if it is too far away; for example, if it is inserted in a cabinet so that one cannot step close to it but must lean forward. The faucet often reduces the usable opening area of the washbasin. Proper height is important.

The toilet is of great importance for elimination of alimentary wastes and for keeping the body clean. Kira's 1976 study provided much information about suitable toilet design, sizing, shaping, height, and location; further recommendations were supplied by McClelland and Ward (1982). Proper handrails and grab bars aid persons with mobility problems, such as caused by back pain, to sit down and get up. Personal hygiene systems installed in toilets are often of great help, as are self-cleaning provisions and any other features that facilitate maintenance.

Kira's publication, as well as that by McClelland and Ward, contains extensive design recommendations for Western-style toilets. Throughout the world, however, different customs and conditions prevail. For these, few ergonomic recommendations seem to have been published (Cai & You, 1998; Ogawa & Arai, 1995).

Bedroom

Most of us stay about one third of the 24-hour day in the bedroom, and weak or ill persons spend even more time in it. Hence, it is important to pay attention to its ergonomic features (H. M. Parsons, 1972). The bed should be at a height that makes lying down and getting up easy. Many different mattress properties have been promoted in the past, ranging from hard over pliant to soft (see the remarks on beds in Chapter 4); when no objective criteria appear sufficient, the user selects the support to please individual preferences. The bedroom must be spacious enough to allow maneuvering space and supply easily reachable shelving and hanging for clothing, linen, and bedding. It should contain communication devices and direct-access storage for medical supplies, and have emergency access as well as an emergency exit. As a rule, the bedroom must provide privacy and be near a bathroom.

DESIGNING NURSING HOMES

In the United States, according to various statistics, only about 25% of 65-year-olds have curtailed functional abilities that cause problems in everyday life. Unfortunately, more serious impairments are likely to appear with increasing age. As new functional disabilities, health deficiencies, or mental problems occur, the aging person first needs some help in his or her own home. Initially, that care may be privately secured either through the spouse or partner, adult children, other relatives and friends, or through hired persons. Close relationships with caring persons can provide the affective, physical, and cognitive intimacy and support that help aging people cope with decreasing self-sufficiency (Roberto, Allen, & Blieszner, 2001; Schulz, Czaja, & Belle, 2001). For some, this is the first step on a path that leads to a nursing home (Birren & Schaie, 2001).

Many aged persons suffer from what Belsky (1990) called *therapeutic nihilism.* They are prone to treatable conditions (possibly self-diagnosed and often wrongly so) but have acquired the attitude of placidly accepting aches, pains, and physical distress that should rightfully be considered a disability (which could be alleviated) as something natural in later life. They may not want to bother their caregiver and not wish to appear weak and discouraged even if they are truly ill, in pain, and in need of aid (Roberto, 2001; Roberto & Gold, 2001). This denial makes it particularly difficult to provide them with the help and care that they need and deserve.

A large number of age-related health disorders are of a musculoskeletal nature, with degenerative osteoarthritic joint disease and associated pain most frequent in North America. Immunological, neurological, and psychiatric conditions are common as well (Birren & Schaie, 2001). Aged people are the most frequent consumers of physicians' services. The normal physician's job is to diagnose an illness, make a medical intervention, and find a cure. However, this is mostly not the case with the older person. The older one gets, the more likely it is that one suffers from a chronic illness that cannot be cured, although occasionally it can be alleviated or covered up, at least for a while. Fighting disability requires diagnostic and care strategies different from battling illness. These include techniques outside the physician's traditional realm of expertise. Physicians who specialize in geriatric medicine collaborate with nurses, physical therapists, dietitians, psychologists, and ergonomists.

There are different kinds of institutions for older persons who cannot stay at home. Some cater to certain religious or ethnic groups, some freely accept people with Alzheimer's disease or those who are bedridden, whereas others do not admit individuals who are severely physically or mentally impaired. Some simply provide room, board, and personal care to their residents; others are more similar to hospitals, offering intensive medical services to seriously ill people. In the United States, the classification of nursing homes depends on the intensity of help that they offer. *Assisted-living* facilities (in the past usually called *intermediate care* homes) are a housing option for older adults who cannot stay in their own homes anymore. *Skilled-care* facilities provide care that is more complete; their services may qualify for reimbursement by Medicaid or Medicare.

For assisted-living facilities, the architectural and other ergonomic recommendations mentioned earlier for the private home also apply because they facilitate residents' efforts to look after themselves. However, further design features and organizational measures lighten the caregivers' tasks, especially in facilitating easy access for cleaning, and can enhance the awareness of a resident's need for help and emergency requirements.

Although it is important that aged people have as much personal freedom as possible, being in an institution limits their choices in even the most basic aspects of life, such as where to live, when to get up or lie down, what to do, and what meals to have. Home management should carefully provide various choices for the residents, keeping in mind their interests, not primarily organizational ease.

The selection of a home can be a difficult task, both in regard to overcoming the emotions involved and in selecting the care needed. In 1990, Belsky stated that the quality of nursing facilities ranged from "home" to "snake pit" (Belsky, 1990, p. 107). Unfortunately, that variation in care still exists today. The Association of Retired Persons reported in its AARP Bulletin of September 2004 on a number of severe problems found in homes for the aged. Most issues, although often related, fell into three categories:

1. *The facility should be clean, organized, inviting, and pleasant to live in.* The AARP report mentioned facilities that were not only unhygienic but just plain dirty, with filthy bathrooms and soiled beds given to newcomers. Meals were "cold, gray, flavorless" so that clients would not eat and lose weight; drinking water was not supplied, so that patients would become dehydrated; and personnel did not assist people who needed help, so that they had to call 911 to have police come to give them a hand.

2. *The facility should be safe, well supervised, and conscientiously run.* Yet, the AARP report stated that many buildings do not have fire alarms or sprinklers. The report also cited cases of Alzheimer's patients not being observed when they should have been supervised; hundreds of registered sex offenders living in nursing homes; one owner of several nursing homes living lavishly from the proceeds that he skimmed off his facilities, thus

causing shortage of soap, disinfectants, bandages, and other basics. He pleaded guilty to federal charges of health-care fraud, inadequate care and services.

3. *The facility should have appropriate medical care.* The AARP report referred to several so-called homes where a responsible physician could spend only about 5 minutes with every client, once a month. Besides unjustifiably high charges for drugs, a large number of cases came to light in which aging patients were severely overmedicated, even wrongly medicated to the extent that their health and well-being were diminished, instead of improved, by the drugs taken.

Selection of a home needs careful consideration; the AARP Web site (http://www.aarp.org) can provide helpful information.

In addition to the clients, there is another group of people of great concern: all the care providers — nurses, caregivers of many specialties, physicians, cleaning and maintenance personnel, and others who, directly or indirectly, serve the people in nursing homes. As the Occupational Safety and Health Administration (OSHA) reported in 2003, the nursing home industry is one of North America's fastest-growing industries. It employs approximately 1.8 million people in about 21,000 work sites; in 2005, the employment level is expected to rise to 2.4 million workers.

Unfortunately, the nursing home industry ranks third highest in nonfatal occupational injuries and illnesses among all U.S. industries. Its has an injury incident rate of 13.9 injuries and illnesses for 100 full-time workers, more than double the incident rate for industry as a whole. Spurred by concerns for the safety of caregivers, in 2003, OSHA issued a set of guidelines targeting the nursing home industry in order to voluntarily reduce injuries.

Regarding the architecture and interior design of nursing homes for people who need intensive care, some of the earlier recommendations for ergonomic designs still apply, but the necessities of providing 24-hr supervision and care, and possibly intensive medical treatment, generate the most important design principles. New architecture and interior design guidelines now incorporate ergonomics information for care facilities (Baucom, 1996; Brummett, 1996; Committee on an Aging Society, 1988; Czaja, 1990, 2001b; OSHA, 2003; Regnier, 1993; Regnier, Hamilton, & Yatabe, 1995). Human engineering of intensive care facilities, such as nursing homes or hospitals, is entirely feasible so that it satisfies the caregivers' needs in providing care and the residents' desires for an appealing homelike environment.

REMODELING AN EXISTING HABITAT

President John F. Kennedy is credited with saying that the time to repair the roof is when the sun is shining. Remodeling an existing home to make it suitable for use in an owner's old age can be a complex task that should be planned and undertaken before the need is imminent. The first step is to establish which redesigns are necessary: for this, the foregoing listing on homes suitable for older persons provides guidance. Multistory residences may need an elevator, the installation of which can be easy and inexpensive if planned during the design of the house. The overall layout of the house may have to be altered to allow for more suitable locations of rooms and better passageways, as discussed, for example, by Bakker (1997), Baucom (1996), and S. Winter (1997). The reworking of the existing residence can be substantial and costly, as Richard Atcheson described in vivid detail.

Home for my wife Jean and I is a nice old two-story Dutch Colonial house in Princeton Junction, New Jersey, built in 1929 with pretty touches — little stained glass windows, for example, and a graceful archway between the living room and the dining room. Hardwood floors throughout. Pleasant neighborhood, good neighbors, all embowered in beautiful old trees and ivy and birdsong.

It is not a fancy house in any way. In fact it's small, kind of shabby and comfortable, all we need or want these days. We moved a lot in the past almost 40 years, a gypsy pair of writers and editors, and have lived in grander places. The house most loved and fondly remembered by our whole family, because we lived there when the children were little and we were young and very, very happy, later became the township hall of Plainsboro, New Jersey, and has since become a museum of township artifacts dating from the 18th century. That generous clapboard house, white with green shutters, sat, in our time, out in the fields, when there were fields, at the end of an avenue of maples. At night when all the lights were on it looked like a ship in full sail. It had a center hall, nine bedrooms, and gracious company rooms that were meant for house parties and dancing and dressing up, and believe me, many a weekend we and our friends and the kids would don our funky finery (it was, after all, the late '60s), put on some bouzouki music, and dance the night away.

But that was way back then. The kids are grown now, and out on their own in New York and London and San Francisco. Incredibly, Jean is 71 as I speak, and I am 66. She works four days a week for the National Football League, commuting by train from PJ (Princeton Junction) to her job in Manhattan. Me, I have an apartment in Washington, D.C., because my work is here. And I chug up to PJ on Amtrak as often as I can for weekends … less often than I'd like.

The house in PJ is not a house that either of us just love. It is too small to suit my persistent grandiosity; I would vastly prefer to dwell in astringent elegance on Library Place in Princeton proper, which we could not ever conceivably afford. But Jean would never hear of such a thing, even if we could; while she will admit to being merely "fond" of our current place, her practical nature tells her (and then she tells me) that it couldn't be more ideal for this time in our lives. It's easy to keep, especially as we employ the services of the Ukrainian ladies, recent immigrants who clean houses while they perfect their English. Two healthy boys come around to mow the little bit of lawn and crop back the verdant Jersey jungle we live in. The small rooms do have a certain cozy charm. And it certainly could not be more convenient for a New York commute, situated as it is practically on the tracks of the Northeast Corridor line, a five minute walk from the PJ train station. And that's really why we bought it in the first place, 16 years ago.

We parcel out various parts of the house in our affections. I absolutely adore the glassed-in front porch because the eyebrow windows on three sides remind me of Dr. Zhivago, and when I open them into the stand of spruce and juniper in front, it makes me feel that I'm in a forest-sweet illusion. I keep my computer out there and play on it in all weathers. We both only tolerate the living room, but actually quite like the dining room, which doubles as a library and houses Jean's computer. And Jean will express real enthusiasm for the rickety back porch, where she likes to sit on summer afternoons to work on other people's manuscripts. We jointly loathe the basement, which is dank and clammy in every season, and piled full of the detritus of our past in musty cardboard boxes and trunks, which neither of us can bear to contemplate sorting out. Oddly, it is the domain of our elder cat, Claus, a grumpy New Yorker guy who rejects society except at dinnertime. Our younger kitty, Emma, loves to snuggle, and will nestle with Jean in the evenings while she is slumbering in front of the TV set, or when we are in bed. Our big mutt Rosie, galumphing part Husky, all love, likes the same intimacy. And so does Ralph, our ancient terrier, who used to be disagreeable and yappy but in old age has mellowed into a true companion; he has certainly kept his looks, but is too creaky to get up on the bed anymore. That's us and that's the resident family, now and so many a time when the house is sleeping, and a light rain is pattering on the roof of the porch and dripping in the trees, I sit out there and feel real peace.

Which brings me to the point, which is our desire and intention to continue living there into our late years, for as long as we can. To be perfectly blunt I guess there will come a time when we don't work anymore, and are "retired," though I can scarcely imagine it. But I got a wake-up call this year when I went to New York City to spend a day making rounds with Rosemary Bakker, research associate of gerontological design and medicine at Weill Medical College of Cornell University, who wrote the book *Elder-design* (Penguin USA, 1997). Rosemary shows people how to make their own homes safe for themselves well into late age. The experience made me realize, profoundly, that Jean and I are well-established members of the over-60 club, and I haven't been able to shut up about this discovery since. My colleagues laugh behind my back. "Good Lord," they have reportedly said, "the man is 66 years old and just realized it?" Well, yes.

I always thought being a writer was so great because I would never have to retire, little thinking then, as a kid, that I would become neither famous nor affluent, and that it would be a question of "tote that barge" forever, like it or not. But actually I like it fine, and Jean feels the same. Still, though we are healthy and quick at the moment, that can't be forever, and we have to think about that, and plan. In aid of which we asked Rosemary if she would come out from New York and look at our situation with a view to the long term. We really do not want to have to go to any sort of "residence for senior living." And I imagined myself in the fantasy scene out of Dickens, in an awful dream about "the life to come," where I am discovered at Rosemary's knees, pleading — "Oh Spirit, tell me that such a thing does not have to come to pass. Tell me that I can change it." And that is exactly what she did.

Rosemary arrived by train one late sunny morning. I met her on the platform and brought her across the station platform and parking lot to our house. I had already observed her ways of sharp observation and gentle counsel; she had become my heroine. But this was the first time for Jean and Rosemary, and I am happy to say that they bonded almost at once. Nor had I been worried about that particularly — Jean is a very welcoming, outgoing person, and was up for the encounter, and Rosemary is the soul of tact, and one of the most compassionate people I have ever known. But there is always a little tension involved when a stranger arrives not just to visit but to go through your house with a fine-toothed comb.

Because it was such a pretty day we had lunch out on our rickety back porch, and after the niceties and getting-to-know-yous, and a sip of Chardonnay, the porch became first subject for discussion, over cold chicken and baby new potatoes and ratatouille. Rosemary pleased me much, saying that being out back gave her the magic feeling that she was miles and miles away in the country … which is what we also feel. However, she did have issues. Number one, as she quickly noted, the porch is too small … small in the sense that a guest, while "skinnying" around the table to a chair, might easily fall off at the top of the steps and plunge into the garden. "And you don't want steps like that," Rosemary observed, "because they really aren't safe." The steps are just graduated planks with no backing. "Somebody could get their leg stuck through there," she proposed.

This horrifying thought led us to an agreement that the whole damn porch has got to go, and the idea of replacing it led to an investigation of what we call "the back kitchen," from which one reaches the porch from inside, which Rosemary noticed was on a level six inches lower than all the rest of the ground floor. And if we had any thought of one day living entirely on one level, it meant that the floor in the back kitchen would have to be raised, which meant that the back door would have to be raised, which meant that the roof above it would have to be raised. Which meant plenty bucks. Jean shot me a penetrating glance and I shot her one right back.

I was not entirely appalled. I have always wanted a bigger back porch. I have always deplored the step down into the back kitchen, and, as Rosemary promptly noticed, the step back up again is a great thing for people to trip on … which she demonstrated personally, showing also that the nearest thing to grab while tripping is the door handle on the refrigerator, which of course swung wide open and flung Rosemary well into the center of the kitchen.

"And Dick," Rosemary continued, "say, if it should happen in 20 years or so that you or Jean was in a wheelchair, one pushing the other, and you hadn't raised the floor here but instead had installed a ramp, which you could do, well, if you were over 80 would you be able to push that wheelchair up?" In my mind I was already talking to a local contractor about raising the roof and the floor and the door and all.

"Or," she went on, "you could just build a new broader, wider porch at this level and think about doing the rest of it later on, when you actually need it." I inwardly whimpered with relief, and the glance between Jean and me this time was much softer.

From the back kitchen we stepped up into the kitchen itself, which we are proud of, having just spent *beaucoup de bucks* on all new appliances and a new floor; it is in the process of redecoration. Rosemary commended us for that. "So many people won't do that," she said. "They keep old things when they don't even work properly anymore. I know people who have lamps that are 50 years old, and they'll say, 'Oh, I can turn this on if I just jiggle it a little.'"

I immediately thought of the delicate china lamp I have in the living room of the apartment in Washington. It was a wedding present to us, almost 40 years old, and it lights only when I jiggle the chain a little. And would defy a human being on Earth to do it.

So after we paused in the kitchen, Jean picked up a spiral pad and a pen for taking notes, and the three of us rambled all through the house, upstairs and down, to the basement and back, and out the front. We got good marks for having wide unencumbered "avenues" of space for walking in rooms and from room to room; Rosemary is fierce about this. But we got some bad marks, too. Yes, it's true that we have to put railings on the basement stairs, lest one of us break his neck down there one day. Yes, it's true that we have little Oriental carpets here and there that might be labeled "fall down and break your hip here." Rosemary's solution: Tape them to the floor. And it is true that the living room furniture is too plump and soft and deep for any old person to get out of without considerable struggle. Rosemary felt that these pieces should be replaced.

Yes, it is true that the front steps, which are made of brick, have no railings. Gotta get those railings. Yes, it's true that the dense clump of trees in front of the house makes it very dark out there at night, and so yes, we need to install more lights ... not only so that visitors can see their way but so that the bad guy will not find it easy to lurk out there. Busted.

[...] But I will speak here of a couple of indelible moments during her visit. One was when we had been up in the bedrooms and upstairs bath and were coming down the stairs again, and Rosemary reached for the Balinese carved figure of Hanuman, the divine monkey king, that sits like a newel post at the bottom of the railing, and it came away in her hand. (Jean, weakly: "Oh, I've been meaning to glue that down again.")

And then there was the moment in the laundry room when we were about to enter the downstairs bathroom (a full bath, luckily for us). Rosemary stopped me and invited me to consider whether I could make that turn if I were in a wheelchair. "You would need 36 inches across," she said, and in my mind's eye I was 36 inches wide. "Oh," she cried, looking into the bathroom. "What is that great big cabinet there? Does it have to be there? You'd never be able to swing into the room." That cabinet is where we keep towels and linen, and I thought, Well, I guess we can keep them somewhere else.

As I wheeled into the bathroom we faced another impediment — the space between the sink and the tub is too narrow to enable me to wheel through to the toilet. "That can be fixed," Rosemary said. "We just need to get rid of that sink and install a wall-hanging one that's not so deep." I imagined myself wheeling through to the toilet. "You'll need some handholds here," she said. I imagined myself lifting my body out of the wheelchair and onto the toilet. "And now here, Dick," Rosemary said, gesturing to the bathtub, "if for any reason — due to arthritis, you know, many people can't climb stairs or step into a bathtub — you could install a hydraulic chair in the tub and with the press of a button it would lower you down into the water." My imagination turned right off. "Or if you wanted to spend the money," Rosemary said, "you could rip out the tub, install a standing open shower, and put a drain in the center of the floor."

After Rosemary left, Jean and I discussed, among many possibilities, doing something about the bathroom. "I think it would be great to have it all wide open," I said, opening my arms expansively.

"You would think that, "Jean said. "But we're not going to have a bath without a curtain. The water would splash all over the floor. And who do you think would mop it up? It wouldn't be you."

"Okay. Whatever we need, somehow we'll find the money."*

SOURCES OF INFORMATION AND ADVICE

The U.S. Census Bureau Web site (http://www.census.gov/hhes) has a section on statistics of older persons.

Many universities have departments that can provide information and help in matters of ergonomic means to assist aging persons; especially departments of gerontology and architecture, engineering (particularly human factors, industrial, and systems engineering), psychology, and physiology may be able to provide specialized advice. Examples of such resources are the Center for Rehabilitation Technology at the Georgia Institute of Technology (www.gatech.edu), the Center for Universal Design at North Carolina State University (www.design.ncsu.edu/cud), and the Trace R&D Center, University of Wisconsin (www.trace.wisc.edu). Local and regional agencies and government offices can be helpful as well. The Access Board, a federal agency, maintains an instructive Web site (www.access-board.gov). The American Association of Retired Persons (www.aarp.org) can provide listings, addresses, literature, and advice. OSHA (www.osha.gov) established guidelines for the nursing home industry.

Design for older persons follows the general guidelines listed in Table 6.2. Obviously, these design principles apply generally to young and old because they make life easy and safe for everybody.

SUMMARY

Our own aging and that of friends and relatives is of perpetual interest and concern. Yet, even today, there exists no comprehensive set of knowledge about changes with age. Anecdotal observations and case studies abound, but available research findings do not provide a complete picture. This is largely so because cross-sectional research is not appropriate to understand the aging process. In most studies, younger adults appear in 5-year age brackets, whereas often those 65 years or older are lumped together as one group, regardless of differences that may exist among them precisely in those attributes that are the object of the research. Clearly, chronological age is not a meaningful classifier. Thus, a basic research task is to establish a suitable reference system with proper scales and anchoring points. This will enable researchers to perform the long-term studies needed to chronicle and understand the aging

---

* From "Our Old House," by Richard Acheson, *Modern Maturity* (November–December 2000 Issue 43R: 6), pp. 62–71. Copyright 2000 by American Association of Retired Persons. Reprinted with permission.

TABLE 6.2
General Ergonomic Goals When Designing for the Elderly

Easy to do
Easy to reach, easy to see
No-bend, no-stretch
Simple maintenance and cleaning
Floor space for easy ambulation, passage, and activities
Non-slip floors, even when wet
No-threshold entries into rooms, shower stalls, closets
Hand rails and grab bars
Toe and knee space for close access to washbasin, cabinet, work surface
Point-of-use storage
Controls of doors, windows, cabinets, appliances: effortless, consistent, common sense, secure
Lighting of common spaces: bright but without glare, uniform
Suitable indoor climate, automatically maintained
Complex issues solved and hidden within the technical system, so that its use is easy and intuitive
Design for human dignity, safety, and comfort

process and its effects on the ability to perform the activities that make up our day, and to ascertain ergonomic help as needed.

It is obvious that many, if not all, physical, perceptual, cognitive, and decision-making capabilities decline with age. Yet, some of those losses are slow and difficult to observe. Other capabilities and faculties decline fast, or they may deteriorate slowly at first and quickly at some point, perhaps stabilizing for a while. Some of these changes are independent of each other, but many are linked, directly or indirectly; failing physical health may have effects on attitude, failing eyesight might lead to a fall and serious injury with ensuing illness. Reviews of the available information indicate the following:

- Changes among aging people do not simply correlate with age in a continuous linear decline but show a variety of rates.
- Variance among aged persons tends to increase with age.
- Changes in function may affect people differently.
- The rate of change can proceed along some dimensions relatively independently of variation in others; yet, in many cases, a serious loss of functional competence in one area accelerates the decline in other capacities.

In 2004, Fisk and his coauthors made the point that common myths imply that older people are less productive than younger persons, that they are less able, less interested in work, and less willing to learn new skills. However, reality is that the current generation of older adults is more diverse, healthier, and better educated than previous generations. Many in this new group of older persons are interested in remaining engaged in the progress of technology, often in some form of productive work, even in starting new careers. These persons are able and willing to learn new skills and take up new tasks.

Ergonomics can supply suitable environments, practices, and tools to counteract many of the declining capabilities of aging persons. Human factors design considers the abilities and limitations that come with aging and helps to create the best possible person-environment fit. Proper design can help all persons of all age groups but is of particular importance for the older individual.

7 Design for Expectant Mothers

OVERVIEW

The expectant mother experiences increases in trunk dimensions and associated difficulties in performing movements. Strenuous physical work, lifting objects, and bending forward all become hard to do, and standing for long periods becomes almost impossible. Furthermore, fatigue is often present, and it may be difficult to maintain attention over extended periods. Most of these problems are especially pronounced during the second and third trimesters. Fortunately, sufficient reliable information is available in the literature to help establish recommendations for the ergonomic design of tasks and equipment suitable for pregnant women.

SPECIAL DESIGNS FOR WOMEN?

Most pregnant women are part of the population of normative adults that ergonomics texts deal with and about whom the first chapters of this book provide information. However, for several months of their lives, they differ from that reference population: Their body dimensions and physical and psychological traits change, affecting some performance capabilities. Capacity for physical work diminishes with the progress of the pregnancy, but after the child's birth, it usually returns to about the previous level, as does the appearance of the body.

Before I address the changes that come with pregnancy, it is appropriate to discuss briefly the general question of whether inherent differences between male and female adults should have ergonomic consequences for the design of equipment and tasks. My earlier publications contain support and references for the following statements — see Kroemer (1997a, 1997b) and Kroemer et al. (2001).

The tables in the Appendix to this book present the exact numbers to support the common knowledge that, on average, women are smaller than men in all anthropometric measurements except in hip width. Yet, as a glance at the standard deviations shows, the diversities in body build (and in power) *within* the gender subgroups are large compared with the variances *between* women and men. Thus, the data distributions describing males and females overlap widely — there are hefty women and feeble men. Furthermore, on the job and in everyday life, the requirements for strength and power are generally well below the maximal exertions; hence, nearly everybody, male or female, can meet task requirements.

Mobility in body joints is generally greater in women, but only marginally so — see Table 4.1 in Chapter 4 for detailed information. On average, girls and women are somewhat more skillful than boys and men on perceptual and psychomotor tests such as color perception, aiming and dotting, finger dexterity, inverted alphabet printing, and card sorting. As a rule, males perform slightly better in speed-related tasks, on the rotary pursuit apparatus, and in other simple rhythmic eye-hand skills.

Girls and women commonly have lower absolute auditory thresholds for pure tones than boys and men. Aging men generally have larger hearing losses than women of the same age. In vision, both static and dynamic, boys and men usually have better acuity than girls and women. With increasing age, females tend to have more vision deficiencies; acuity, especially, declines earlier in females than in males. In gustation (taste), many women may detect sweet, sour, salty, and bitter stimuli at lower concentrations than men can. In olfaction (smell), women can usually detect some substances more easily than men can. Both taste and smell capabilities and preferences vary within the female menstrual cycle and during pregnancy.

In the cutaneous senses, the genders exhibit little difference in the thresholds for temperature sensation. Males may find it slightly easier to adjust to very hot and very cold conditions. Many females feel initially less warm (comfortable) in cool environments but adapt to the surroundings more rapidly than males. Females often begin to sweat at higher temperatures than males do and get acclimatized to working in severely hot conditions somewhat more slowly. Regarding touch, the sensitivity to vibration is about the same in males and females. However, females are more sensitive to pressure stimuli on their bodies, except on the nose. Pain sensations (a complex and difficult topic for physiological and behavioral reasons) seem to be about the same in males and females.

The popular concept of "stress" encompasses a wide variety of situations, including job pressures, health concerns, marital and family tensions, and physical aspects of the environment such as climate and noise. Among these, few gender-specific differences in stress tolerance have been demonstrated. Whereas females may be slightly more sensitive to sound, the impact of noise on performance, health, and social behavior seems to be similar for both sexes. Women appear to be better able to cope with less personal space than men; when forced to be in crowded quarters, men tend to maintain greater distances from others, and they also react more negatively when others invade their personal space. As a group, women may be able to cope better with the monotony and boredom of low-arousal conditions of work, whereas men might deal better with the pressure of higher arousal conditions.

Performance is another topic of general interest. Task performance depends on various attributes, including the overall nature of the chore, the specific conditions under which the task at hand is performed, the requirements for special capabilities, the attitudes of the person performing the assignment, the subjective value of the task for the person performing it, and the social and task-related goals of the individual. Kroemer et al. (2001) cited a number of studies that found, under comparable circumstances, no differences at all, no consistent differences, or only minor differences in the performance of women vis-à-vis men. Even in areas where differences — usually small — have been found, such as in reactions to crowded conditions or in fine manual skills, it is uncertain whether they reflect basic biological or genetic differences or are merely or mostly the results of gender roles acquired through group traditions under social pressure in certain cultural contexts.

In conclusion, with respect to equipment and task design, interindividual variability overshadows by far any existing differences between the groups of so-called regular adult men and women. Thus, there are no gender-specific traits that require design of workstations or work tools exclusively for either men or women (except for expectant mothers, as will be discussed in some detail). Proper adjustment ranges can accommodate most people, whether female or male. Nevertheless, one group or the other is likely to prefer certain jobs, although probably not exclusively so; for example, strong men do most tasks that require brute force, but not all men are strong.

CHANGES IN BODY DIMENSIONS OF EXPECTANT MOTHERS

As just discussed, in general, there is no need to design work, work equipment (besides tight-fitting clothes), and workplaces specifically for either women or men — with one exception: Expectant mothers do require particular ergonomic consideration because of temporary changes in body dimensions and capacities, especially for physical work. Changes in body dimensions with pregnancy become apparent after two or three months' gestation, and most size measurements increase throughout the course of pregnancy. The most obvious expansions are in protrusion and circumference of the abdomen and in body weight.

A review of the existing anthropometric, biomechanical, and other human factors information on pregnant women leads to the surprising and disturbing conclusion that many descriptions of the changing body still seem to be based more on everyday experience and small samples of convenience than on recent, large-scale, well-planned, and comprehensive surveys. Pheasant (1986) attempted to compile anthropometric data on expectant mothers in the United Kingdom. He was able to find only one sufficiently recent anthropometric survey of pregnant women, but that had been conducted in Japan. Thus, he estimated the anthropometric changes of British women from Japanese data.

Fluegel, Greil, and Sommer (1986) reported the changes from the fourth month of pregnancy as measured on 198 German women. Just before giving birth, on average their weight had increased by 17%, waist circumference by 27%, chest circumference by 6%, and hip circumference by 4%. For 105 white American women with an average age of 26 years, Rutter, Haager, Daigle, Smith, McFarland, and Kelsey (1984) reported the following average increases, seen just before birth, above the values measured 16 weeks after the onset of pregnancy: 17% in body weight, 8% in chest circumference, and 4% in hip circumference. In a 1997/8 U.S. Air Force study, Perkins and Blackwell (1998) reported that, in the final stage of pregnancy, body weight had increased by 21% over the value 8 weeks into pregnancy, waist circumference enlarged by 35%, chest circumference by 10%, and hip circumference by 7%.

That survey of pregnant women by Perkins and Blackwell had a small sample size (15 women) but nevertheless deserves particular attention because it is somewhat recent and planned, executed, and described especially well. Furthermore, it is of great interest because it reported on two different kinds of measurements:

1. traditional point-to-point distances that were measured with the "classic instruments"; that is, anthropometer, caliper, tape, and so forth
2. three-dimensional surface-contour dimensions that were obtained with a whole-body laser scanner.

(Chapters 2 and 4 contain discussions of anthropometric measurement techniques.)

The results of the traditional measurements were processed as usual and reported primarily as percentage changes, and the surface-contour measurements were processed in the form of radial difference maps and euclidean distance matrix analyses.

The participants were

- between 23 and 37 years of age, almost all Caucasian
- measured six times: once during the first trimester (before any physical changes were observable, on average after 8 weeks of pregnancy), once during the second trimester (on average after 21 weeks), thrice during the third trimester (on average after 28, 32, and 37 weeks), and once after delivery (on average after four weeks).

Measurements began during May 1997 with 35 women. Ten participated only in the first measurement gathering, whereas 15 completed all 6 sessions. The measurements were conducted with the women placed as close as possible in the standardized erect standing and sitting postures (described in Chapter 4).

The results showed, as usual, a wide variety in body changes among the women. Body shape and mass properties remained fairly constant during the first three months of pregnancy, but then the fetus' growth, apparent in the increases in the mother's body weight and abdominal depth and circumference, brought with it many other anthropometric changes. The researchers recorded widening of the pelvic girdle as evidenced by a larger hip width, backward sway of the trunk, spread and elevation of the lower ribs with an increase in chest depth, and growth in the vertical distance between the pelvis and the lowest ribs. So, the expansion of the expectant mother's lower torso is actually in all directions, although predominantly forward. Figure 7.1 shows the typical development of changes in body contours, as seen from the right side.

These physical changes in body size, contour, amount and distribution of mass, as well in posture, bring about a biomechanically significant shift in the center of mass of the trunk. This affects the pregnant woman's gait (Foti, Davids, & Bagley, 2000) and her efforts in doing everyday activities, and specifically raises concerns about protection against injury, especially in an automobile accident as either driver or passenger.

Culver and Viano (1990) depicted these changes as ellipses, which allowed estimates of the contact area between the woman's body and restraining devices or the interior surfaces of automobiles in the case of a crash impact. Klinich, Schneider, Eby, Rupp, and Pearlman (1999), Klinich, Schneider, Moore, and Pearlman (2000),

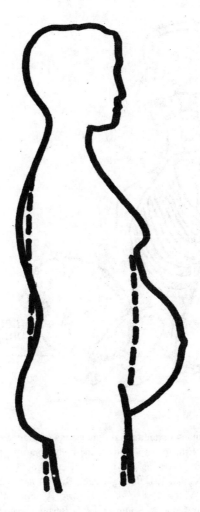

FIGURE 7.1 Typical body contour and posture of an expectant mother. Note the changes in back curvature and general pose due to pregnancy. Adapted from *Anthropologischer Atlas,* by B. Fluegel, H. Greil, and K. Sommer, 1986, Berlin, Germany: Tribuene.

and Schneider et al. (2000) gathered new information that they applied to the protection of pregnant women and their fetuses in the case of a vehicle crash. Their work included developing state-of-the-art biomechanical models of the pregnant body and analyzing its interaction with automobile restraint systems, including air bags. Figure 7.2 shows a typical model of an expectant mother sitting on a car seat.

Using Seat Belts

Personal experience and anecdotal evidence indicate that many pregnant women hesitate to use seat belts in cars because they fear that in an accident, the restraining

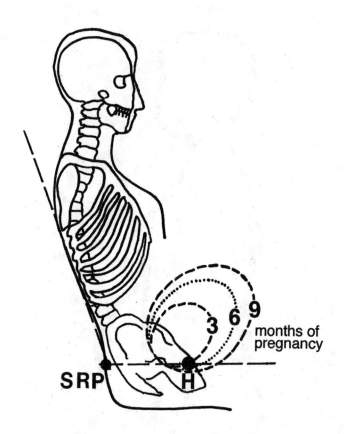

FIGURE 7.2 Typical changes in abdominal dimensions of pregnant women when sitting in a typical automobile seat. Ellipses approximate the body contours. SRP is the seat reference point, and H the hip point, according to SAE practices. (Adapted from "Anthropometry of Seated Women During Pregnancy. Defining a Fetal Region for Crash Protection Research," by C. C. Culver and D. C. Viano, 1990, *Human Factors, 32*, pp. 625–636; *Anthropologischer Atlas* by B. Fluegel, H. Greil, and K. Sommer, 1986, Berlin, Germany: Tribuene; and *Seated Anthropometry During Pregnancy*, by K.D. Klinich, L.W. Schneider, B. Eby, J. Rupp, and M. D. Pearlman, 1999, Ann Arbor, MI: University of Michigan, Transportation Research Institute.)

device might be harmful instead of helpful. Here are the personal opinions of Kathleen Klinich, research engineer with the University of Michigan's Traffic Research Institute:

> Our crash investigation program found that unrestrained or improperly restrained pregnant women were 5.7 times more likely to have an adverse fetal outcome than properly restrained pregnant women. Women restrained by 3-point belts and air bags seemed to have slightly better fetal outcomes than those restrained only by 3-point belts. All of the severe crashes (greater than 30 mph delta V) had adverse fetal outcomes regardless of restraint use, but belt use had a positive effect in minor and moderate severity crashes, which make up over 95% of all crashes. Overall, our research supports

recommendations by NHTSA [National Highway Traffic Safety Agency] and the American College of Obstetrics and Gynecologists' recommendations to properly wear a three-point belt and not to disconnect the air bag.

So what does this mean for a pregnant occupant? Having been pregnant twice over the course of this research, these are the steps I took when in a car.

1. Always wear the three-point belt, with the lap belt positioned as low as possible and the shoulder belt routed over the sternum.
2. Leave the air bag operational.
3. Drive the family vehicle that allows the most clearance between the steering wheel and my abdomen. If the distance[s] were about the same in our vehicles, I would choose the newer vehicle, which theoretically should have more advanced safety features.
4. When possible, ride as a passenger rather than drive (no steering wheel to contact).
5. Remove or adjust coats to make sure they don't interfere with low placement of the lap belt.
6. Minimize travel in hazardous driving conditions.
7. Position steering wheel to maximize abdomen-to-wheel clearance, while aiming it more towards the chest than the head.

Some of these steps are based on my personal opinion of best practice for pregnant women and are not necessarily supported in a statistically significant way by our research. For example, our study of crashes did not find a difference in fetal outcome for drivers and passengers, but since we had a few cases where the steering wheel contributed to adverse fetal outcome, it seems reasonable to try to avoid steering-wheel contact by riding as a passenger where possible." (Reprinted with Ms. Klinich's permission from her personal communication of August 8, 2001, to the author.)

CHANGES IN WORK CAPACITY
DURING PREGNANCY

The growing abdominal protrusion makes it increasingly difficult for a pregnant woman to bend forward and to get as close to work objects as she could when not pregnant. The available working area of the hands in front of the body gets smaller during pregnancy. Manipulating objects that are now farther ahead of the spinal column generates an increased compression and bending strain on the spine and on ligaments and muscles in the back. This loading is also due to the increasing mass of the abdomen, and its expanding moment arm with respect to the spinal column. The increasing bulkiness of the abdomen changes the body posture, which, in the course of the pregnancy, shows a backward pelvic rotation, accompanied by forward movement of the trochanterion and a backward shift of the upper trunk. This could cause a flattening of the lumbar bend, but Perkins and Blackwell (1998) recorded a "swayback," an accentuated lordosis, in their participants. This explains, at least partly, the complaints of back strain and back pain common in pregnancy (Cherry, 1987; Fast, Shapiro, & Edmond, 1987). Engaging in specific exercises to strengthen the abdominal and back muscles may relieve backaches in pregnancy. Some of these exercises can be performed in work environments.

Physical performance capabilities change during pregnancy, but there are great variations among individuals. By the beginning of the ninth month, blood volume typically has risen by up to 40%, together with cardiac output and blood pressure. Maternal metabolism increases throughout pregnancy, usually associated with a warmer body temperature. These events can make working in warm and humid environments difficult. With advancing pregnancy, in general the ability to perform the following types of work decreases:

- work requiring great exertion
- work requiring a great deal of mobility, such as low bending and far reaches
- work involving frequent repetition or work extending over long periods (Cherry, 1987; Errkola, 1976).

Perkins and Blackwell (1998) found that when asked to stand upright for 10 min at a time, some of the pregnant women experienced loss of balance, or when asked to sit upright, were not able to do so for a full 10 min because of back pain. Several women were not able to put their knees together while sitting.

In general, a suitable sitting posture is less strenuous than standing, especially if interrupted by getting up and moving about, or even lying down for a while. Sitting still for prolonged periods during pregnancy is not medically advisable due to restriction of circulation to the lower extremities. Wearing support hose and frequently getting up from the seat to walk can alleviate the risks associated with prolonged sitting during pregnancy. Therefore, many women can continue desk jobs until delivery, whereas jobs that require prolonged standing, forceful lifting, pushing and pulling, climbing of ladders and stairs, bending below waist height, and other intensive and repeated physical efforts need to be lightened or even entirely discontinued during the course of pregnancy. In addition, pregnant women may have to abandon jobs that require frequent changes in work shift assignments (Eisenberg, Murkoff, & Hathaway, 1991).

Nicholls and Grieve (1992) reported on a survey in London of 200 women who were between 29 and 33 weeks pregnant. The women compared their current abilities on 46 tasks with their performance before becoming pregnant. Of the 46 tasks, 32 were significantly more difficult to perform during pregnancy than prior to pregnancy. Among these difficult activities, the following were considered the hardest:

- picking objects up from the floor
- walking upstairs
- driving a car
- getting in and out of a car
- using seat belts in a car
- ironing
- reaching high shelves
- getting in and out of bed
- using public toilets.

Most of the perceived reasons for the difficulties related to back pain and reduced reach and mobility. Owing to the increasing bulk of the trunk with pregnancy, objects on the ground near the feet are hard to see, and stumbles and falls are dangers, especially with an existing loss of balance. Frequent urination is a common symptom of pregnancy. Therefore, convenient access to a restroom is an important workplace feature for the pregnant worker.

Many everyday tasks become more difficult with pregnancy, regardless of the specific requirements and expectations associated with living in different parts of the world — say, Asia (Hirao & Kajiyama, 1994; Hui, Lee, Her, & Tsuang, 1999) or Europe (Paul, Frings-Dresen, Salle, & Rozendal, 1995).

Pregnancy may also affect concentration and information retention, resulting in what some women describe as feeling "scatterbrained." This syndrome is probably caused by hormonal changes that accompany pregnancy. Yet, differences among individuals are striking, with some women finding only a few tasks more arduous, and other women finding nearly all activities harder to do — the more so the closer the delivery is.

Some pregnant women experience widely varying mood states. Although often joked about, affective lability in pregnancy may indeed be disabling. Frequent episodes of weepiness can be embarrassing to a woman at the work site, especially if they are difficult to predict or control. Far more worrisome, and fortunately much less common, is clinical depression related to pregnancy. When depression occurs, it is most likely to begin in the first trimester. Implications for task performance are decreased attention and concentration, impaired cognitive function, psychomotor retardation, decreased initiation of activity, as well as fatigue and apathy. Fortunately, few women feel depression during their pregnancy. In the United States, about 10% of pregnant women undergo mild to moderate depression during pregnancy; severe depression occurs in 1 or 2% of expectant women.

Morris, Toms, Easthope, and Biddulph (1998) reported on the cognitive and mood states of 38 working women in the last third of their pregnancies. Their test results compared with those of a matched control group of nonpregnant female workers showed few significant disparities. The existing differences mostly indicated that pregnant women felt less alert, vigorous, and energetic and were more easily fatigued.

Fatigue in pregnancy, especially after the twentieth week of gestation, may be caused by iron-deficiency anemia. This anemia, which is easily treated, is especially common in women with multiple fetuses, those who have had several babies in quick succession, and those with poor nutrition or frequent vomiting due to pregnancy. There are other common causes for fatigue in pregnancy, including inadequate nutrition, compromised sleep due to discomfort or frequent need for urination, and decreased work capacity. Fatigue tends to be most pronounced in the first and third trimesters, although some (but not all) pregnant women may experience surges in activity and well-being in their second trimester. Commonly, fatigue, discomfort, and pain are more pronounced in the third, fourth, and subsequent pregnancies. This may be caused by the added burden of taking care of other children while pregnant, advancing maternal age, wear and tear on the body, cumulative weight gain, and combinations of these circumstances.

Besides specific advice rendered by a pregnant woman's physician, the time-honored adage applies: listen to what your body tells you: do not do anything that

- hurts
- increases the risk of a fall
- appears unsafe
- feels awkward
- fatigues unduly
- exacerbates any existing problems such as backache or headache.

ERGONOMIC DESIGN RECOMMENDATIONS

To accommodate pregnant women, either at the workplace, in transportation, or at home, Kroemer et al. (2001) suggested the following ergonomic measures:

- Position manipulation areas close to the body and possibly somewhat above their regular height.
- Provide work tasks that require as little force as possible, particularly in vertical directions.
- Avoid lifting of objects of any size or weight.
- Provide suitable seats, easily adjustable by the woman. In general, chairs with a high back, good support, and a firm cushion are preferred. Also, a footrest allowing slight elevation of the legs is often helpful.
- Allow frequent breaks in the work, freely selected by the woman. A special room for expectant mothers to rest, refresh themselves, and use the toilet is highly desirable.
- Provide more space than usual for moving around, and remove obstacles, particularly low objects that might be difficult to see.
- If standing on the job is required for periods, provide some kind of a footstool on which to temporarily rest a foot with the knee bent.
- If the employer provides work shoes, make some adaptation for expectant mothers. Suitable orthotic inserts can offset the distorted center of gravity of advanced pregnancy. Both very high heels and completely flat shoes may be ill advised. Because feet often swell and even grow during pregnancy, expectant women may need wider shoes than those used before pregnancy.

SUMMARY

The expanding abdominal protrusion, increased body mass, and greater circulatory stress accompanying the progress of pregnancy make it increasingly difficult to move about, bend the body, reach distant objects, do lifting, and see obstructions near the feet. Ergonomic measures to accommodate pregnant women at the workplace or at home are fairly straightforward. They relate primarily to keeping manipulation areas close to the body at about elbow height, exerting little force with the hands, and avoiding trunk bending and especially lifting. A suitable seat that can be easily adjusted while seated should be provided. Frequent rest periods should be allowed, freely selected by the expectant mother.

8 Design for Children and Adolescents

OVERVIEW

Children are close to our hearts. They are vulnerable and must be protected but should also be exposed to new experiences and surroundings so that they can learn and develop into well-adjusted and balanced adolescents and, finally, mature adults. Parents provide a loving environment, nourishment, and shelter to the child. But supplying food, furniture, strollers, car seats, toys, and clothes for babies, kids, and teenagers is big business as well. Anthropometric, biomechanical, physiological, and psychological information is available to design environments properly so that they are safe and suitable for the body and mind of a child.

CHILDREN GROW INTO ADOLESCENTS

During our growing years from birth to early adulthood, we undergo major changes in our body dimensions and strengths, cognitive and physical skills, and all other physical and psychological traits. Table 8.1 provides an overview of that development. From infancy through early childhood, ergonomic design mostly concerns the comfort and safety aspects of cribs and clothes, toys and playgrounds, and strollers and car seats. Later, roller skates and bicycles and sports equipment and electronic apparatus appear as human factors challenges, as do computer workstations and school furniture.

Early adolescence poses diverse design challenges because of the huge variety in body size and power and in interests and activities that appear among girls and boys. Some of them remain children longer than others, but eventually all become young adults, most of whom fit into the normative category of "normal people" — see earlier chapters of this book.

Naturally, the responsibilities of caring for children rest with their parents and, to some degree, with teachers. But it is apparent from the abundance of advertisements on television and in newspapers that babies, children, and teenagers are also of great interest to big business. According to an Associated Press story of September 1, 2001, many department stores make huge profits in the sections devoted to kids' merchandise. Some retail chains have stores exclusively directed at children from 7 to 14 years (such as Limited Too in the United States) and for teenagers (for instance, Abercrombie & Fitch, Gadzooks, Inc., Gap, Hot Topic, Inc., JC Penny, Limited Too, and Pacific Sunwear). It appears that the anthropometric, biomechanical, physiological, psychological, and sociological knowledge these merchandisers have about

TABLE 8.1
Development Stages From Infancy to Adolescence

Stage	Physical Characteristics	Motor Skills
Infancy and Toddlerhood		
0–6 months	Oversized head, short stubby limbs	Reaches and grasps, sits with support
6–9 months	Increase in weight, hence plump appearance	Sits alone, stands with assistance
9–15 months	Flexible limbs	Crawls, walks, stands alone
After 18 months	Gradual appearance of neck, protruding abdomen	Wobbly, stiff, flat gait; climbs
2–3 years	Head has become smaller in proportion to the body; less roundness, lean muscles, curved back, still protruding abdomen	Flexible at knees and ankles; can run, jump, kick, hop
Early Childhood		
3–5 years	Rate of growth slows, body proportions change; loss of the baby-like appearance, increase of muscle tissue, less back curvature	Masters walking; has smoother movements, better balance; turns corners; holds pencils and utensils
Middle Childhood		
6–12 years	Horizontal growth, gradual changes in physical appearance	Increase in running and jumping distance, accuracy, and endurance
Adolescence		
12–18 years	Growth spurt peaks, hands and feet reach adult size; breasts develop in girls, body breadth increases; elongated trunks and legs in boys	Motor skills fully developed

these age groups is greater than that appearing in the scientific literature, information perhaps gathered informally and probably protected as trade secrets.

ANTHROPOMETRY OF CHILDREN AND ADOLESCENTS

At birth, we weigh about 3.5 kg and are about 50 cm in length. In the two decades that follow, body length increases three- to four-fold and weight increases about twenty-fold. Body proportions change drastically. In newborns, the trunk accounts for about 70% of stature, and in adults, for just over 50%. Figure 8.1 illustrates the change in proportions over time.

Body size measurements are difficult to take on infants and toddlers because they do not understand or follow instructions. For that reason, up to the age of 2 years, the body length of children is customarily measured while they are lying on their backs. Recumbent length is up to 2 cm longer than standing height (Hamill, Drizd, Johnson, Reed, Roche, & Moore, 1979). Later, the measurement can be taken with

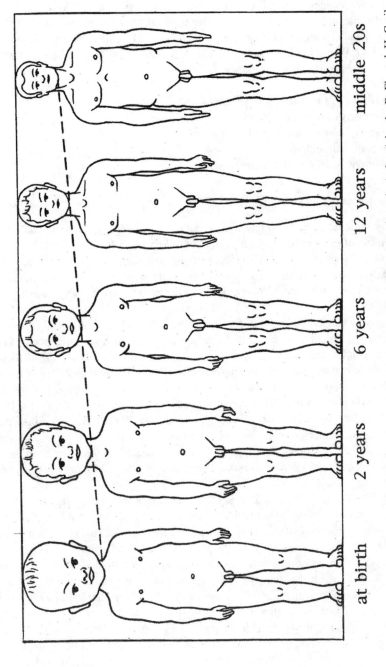

at birth   2 years   6 years   12 years   middle 20s

FIGURE 8.1 Changes in body proportions from birth to adulthood. (Adapted from *Anthropologischer Atlas*, by B. Fluegel, H. Greil, and K. Sommer, 1986, Berlin, Germany: Tribuene.)

the person standing and the same anthropometric techniques can be used as with adults. (Chapters 2 and 4 contain detailed information on common measurements.)

In recent decades, several secular trends have appeared in North America, Europe, Russia, and Asia (Cole, Belizzi, Flegal, & Dietz, 2000; Hamill et al., 1979; Siervogel, Roche, Guo, Mukherjee, & Chumlea, 1991; Tsunawake et al., 1995; Y. Wang, J. Q. Wang, Hesketh, Dang, Mulligan, & Kinra, 2000). In comparison with the observations made about a century earlier

- birth weight and length have increased
- the rate of growth during childhood has increased
- puberty now occurs at earlier ages
- average adult stature has increased
- average body weight has increased.

OVERWEIGHT AND OBESITY

Unhealthy increase in body weight among children and adolescents (and adults) is a recent issue of great concern (Cole et al., 2000; Strauss & Pollack, 2001; Y. Wang et al., 2000). As early as in 1997–1998, school-based surveys of about 29,000 13- to 15-year-old boys and girls in 15 countries showed that body mass indices in many countries indicated widespread overweight or obesity. The worst finding was in the United States.

Commonly attributed reasons for the galloping epidemic of obesity among children and adolescents in the United States are the increasing indulgence in fast foods, soft drinks, and manufactured, prepackaged snacks — nearly all of them having high calorific content but lacking nutritional value (Brownell, 2004). A study of more than 6,000 Americans (S. A. Bowman, Gortmaker, Ebbeling, Pereira, & Ludwig, 2004) showed that on a typical day, nearly one person in three munched fast food and guzzled soft drinks. That supplied almost 200 calories more than the fare contains that a person would otherwise consume. "Supersizing" of snacks and drinks, meals in restaurants, and popcorn in movie theaters fattens youngsters and grownups, so that in 2003, about 60% of American adults and 20% of children were overweight or obese.

Brody (2004) said that in the United States, schools from kindergarten to college contribute to the spread of obesity. Underfinanced schools invite fast-food purveyors and soft drink vendors to their facilities, often expecting kickbacks in money or equipment. The effect is not only access to an enticing food supply high in "junk energy" and low in nutritional value, but also accustoming the kids to that kind of food and drink, a habit that they are likely to continue throughout their lives.

At the same time, physical education programs and athletic activities, both indoors and outdoors, were squeezed out of the school schedule, while physical play and sports yielded to television, video games, and computers. Safe activity centers, swimming pools, exercise tracks, and hiking and biking trails are low on the to-do list of money-starved communities. The overall result is moving less and eating more, thus putting on fat cells, possibly for life.

GROWING BODIES

Changes in body size during childhood vary from individual to individual; they depend on genetic factors, nutrition, health, and activities and environmental variables. Usually, body size increases rapidly during infancy (up to 2 years), and then more slowly until the onset of puberty, but actual growth often occurs in spurts. In most boys, growth accelerates at about 11 years, reaches its peak at about 14 years, and then slows until adult stature is attained in the early to middle 20s. Yet, at 14 or 15 years of age, some boys have reached almost adult size, whereas others are just beginning to enter the rapid-growth phase. In girls, the puberty growth spurt commonly begins earlier, at about 9 years of age, and is fastest around 12 years; full adult stature is often complete at 16 years. Hence, in the age group of 11 to 13 years, many girls are taller than boys of the same age.

Of the anthropometric information compiled in Tables 8.2 and 8.3, data on children in the United Kingdom stem from estimates rather than measurements (Pheasant, 1986). The information on children in the United States, supplied by Hamill et al. (1979), relies on data measured between 1963 and 1975; the numbers were statistically smoothed through spline functions. The U.S. data measured by Snyder, Spencer, Owings, and Schneider (1975) and Snyder, Schneider, Owings, Reynolds, Golomb, and Schork (1977) are also more than 30 years old. The measurements of German children (Fluegel, Greil, & Sommer, 1986) date mostly from the early 1980s. The data on children from France, Japan, and the Netherlands are more recent (Ignazi, Martel, Mollard, & Coblentz, 1996; Tsunawake et al., 1995; Steenbekkers, 1993).

A 1995 report by Norris and Wilson contains older information on children in Belgium, China, Hungary, India, New Zealand, Norway, and Saudi Arabia. Mououdi and Choobineh (1997) listed even earlier surveys done in Algeria, Holland, Korea, Nubia, and Turkey. Roche and Malina (1983) and Malina and Roche (1983) compiled older information on North American children. The U.S. Centers for Disease Control (CDC) continuously collects data an American children (and adults) through the National Health and Examination Survey (NHANES) — see the growth charts in Figures 8.2 through 8.5.

Several observations illuminate the existing state of anthropometric information on children and adolescents. First, there is a dearth of recent comprehensive and representative data on most populations. The extensive compilation of international information by Norris and Wilson (1995) could use only four data sets published in the 1990s, whereas all others were from the 1980s and 1970s. Cole, Belizzi, Flegal, and Dietz (2000) and Y. Wang et al. (2000) mentioned large data sets from Russia, which, however, do not seem to have appeared in the open literature.

Second, among the few recently published sets of anthropometric data, some are obviously not representative of the underlying general population. For instance, nearly 200 children measured in only one school, apparently in Maharashtra (Ray, Ghosh, & Atreya, 1995), cannot represent all children on the Indian subcontinent, where great variability in anthropometry exists (Victor, Nath, & Verma, 2002).

Third, some publications report data in unusual forms: Ray et al. (1995) combined all measurements into unisex tables. Mououdi and Choobineh (1997) combined

TABLE 8.2
Average Stature in cm (With Standard Deviation) of Female Children and Adolescents)

C   China (Taiwan) (Wang et al., 2002b) — measured
F   France (Ignazi et al., 1996) — measured
G   Germany (Fluegel et al., 1986) — measured
J   Japan (Tsunawake et al., 1995) — measured
N   Netherlands (Steenbekkers, 1993) — measured
P   Portugal (Froufe et al., 2002) — measured
S   Sweden (Haeger-Ross & Roesblad., 2002) — measured
UK   United Kingdom (Pheasant, 1986) — calculated
US1   United States (Snyder et al., 1975) — measured
US2   United States (Hamill et al., 1979) — calculated
US3   United States (CDC Growth Charts, 2001); see Figures 8.3 and 8.5

Age (years)	C	F	G	J	N	P	S	UK	US1	US2	US3
Birth	—	—	51.8	—	—	—	—	—	54.8 (3.6)	—	49
0.5	—	—	68.3	—	—	—	—	—	68.6 (2.3)	—	65
1	—	—	75.6	—	—	—	—	—	72.4 (2.9)	—	74
2	—	—	85.9	—	92.9 (4.6)	—	—	89	84.0 (3.4)	86.6	85
3	—	—	94.1	—	100.4 (4.5)	—	—	97	92.9 (4.4)	94.1	94
4	—	107.2 (5.2)	101.3	—	108.2 (4.0)	—	102.7 (1.5)	105	99.5 (4.3)	101.6	101
5	—	110.8 (4.7)	107.2	—	115.9 (4.9)	—	110.3 (4.0)	110	106.5 (4.7)	108.4	108
6	118.4 (5.8)	116.8 (4.9)	115.1	—	122.7 (4.9)	119.8 (5.6)	114.8 (5.8)	116	112.8 (5.0)	114.6	115
7	122.4 (5.8)	125.0 (5.4)	121.0	—	128.6 (5.7)	125.7 (6.5)	123.2 (6.6)	122	118.8 (5.0)	120.6	122
8	127.6 (6.3)	134.0 (5.0)	126.1	—	134.1 (5.4)	130.4 (5.9)	131.1 (6.6)	128	123.4 (5.3)	126.4	128
9	134.4 (7.2)	136.8 (5.7)	130.2	—	139.2 (6.4)	135.5 (6.6)	136.6 (6.8)	133	130.2 (5.9)	132.2	133
10	138.9 (7.3)	142.7 (7.7)	137.2	—	147.1 (6.6)	142.9 (8.7)	138.8 (6.6)	139	134.4 (6.1)	138.3	138
11	146.3 (7.3)	147.8 (6.9)	142.7	145.8 (7.3)	151.0 (6.7)	—	148.0 (9.8)	144	141.1 (6.8)	144.8	144
12	152.7 (6.4)	153.7 (6.2)	148.3	150.2 (5.8)	156.6 (7.7)	—	153.6 (7.5)	150	145.5 (6.5)	151.5	151

TABLE 8.2 (continued)
Average Stature in cm (With Standard Deviation) of Female Children and Adolescents)

Age (years)	C	F	G	J	N	P	S	UK	US1	US2	US3
13	156.5	158.0	154.6	155.1	—	—	158.4	155	155.1	157.1	157
	(5.2)	(4.9)		(4.5)			(7.6)		(6.2)		
14	157.6	160.7	160.0	156.2	—	—	162.4	159	—	160.4	160
	(5.2)	(6.7)		(5.3)			(7.5)				
15	158.2	161.8	162.2	157.2	—	—	168.8	161	—	161.8	162
	(5.3)	(6.7)		(5.2)			(6.3)				
16	157.9	162.4	162.9	158.4	—	—	167.2	162	—	162.4	163
	(4.9)	(6.0)		(6.1)			(5.0)				
17	158.8	162.9	163.5	157.2	—	—	—	162	—	163.1	163
	(5.1)	(5.2)		(5.0)							
18	—	162.8	163.9	157.8	—	—	—	162	—	167.7	163
		(5.5)		(5.7)							

all measurements taken on Iranian children aged 6 to 11 years into one age block. M. J. J. Wang, E. M. Y. Wang, and Lin (2002b) merged results from Taiwan into 3- and 6-year spans, but their original compilation (M. J. J. Wang, E. M. Y. Wang, & Lin, 2002a) of the measurement data listed these in the customary one-year age spans.

Here is a short note for the statistical nitpickers among us: Some anthropometrists categorize children according to their nominal age; that is, they call a child 4 years old from his or her fourth birthday until the fifth birthday, for example. Another listing scheme, however, uses the six months each before and after the actual birthday, from 3.5 to 4.5 years, which means that children thus sorted are, on average, half a year younger than those classified by nominal age. The two groupings yield numerically different anthropometric information, which might be of practical importance for the early fast-growth years.

GROWTH CHARTS

Graphics help to depict the overall changes in body dimensions from childhood to adolescence, but smoothed curves also camouflage spurts and individual differences. For North American children and adolescents, growth charts can be downloaded, at no cost, from a CDC Web (www.cdc.gov/nchs/about/major/nhanes/growthcharts/clinical_charts.htm). The curves shown in Figures 8.2 through 8.5 are those available at the time of printing of this book. The reader may want to visit the CDC Web site to view the most current information.

TABLE 8.3
Average Stature in cm (With Standard Deviation) of Male Children and Adolescents)

C China (Taiwan) (Wang et al., 2002b) — measured
F France (Ignazi et al., 1996) — measured
G Germany (Fluegel et al., 1986) — measured
J Japan (Tsunawake et al., 1995) — measured
N Netherlands (Steenbekkers, 1993) — measured
P Portugal (Froufe et al., 2002) — measured
S Sweden (Haeger-Ross & Roesblad, 2002) — measured
UK United Kingdom (Pheasant, 1986) — calculated
US1 United States (Snyder et al., 1975) — measured
US2 United States (Hamill et al., 1979) — calculated
US3 United States (CDC Growth Charts, 2001); see Figures 8.2 and 8.4

Age (years)	C	F	G	J	N	P	S	UK	US1	US2	US3
Birth		—	52.4	—	—	—	—	—	55.4 (4.0)	—	50
0.5	—	—	69.6	—	—	—	—	—	70.4 (2.4)	—	67
1	—	—	76.4	—	—	—	—	—	73.5 (3.2)	—	75
2	—		86.9	—	93.9 (4.5)	—	—	93	85.3 (3.4)	86.8	87
3	—		95.0	—	102.1 (4.4)	—		99	93.4 (3.9)	94.9	95
4		107.1 (4.6)	102.2	—	108.5 (4.7)		103.4 (6.9)	105	99.9 (3.8)	102.9	102
5	—	111.3 (4.6)	108.1	—	117.0 (4.8)	—	109.0 (4.9)	111	107.6 (5.0)	109.9	109
6	119.3 (5.3)	118.2 (5.8)	116.1	—	122.5 (4.7)	119.2 (6.0)	116.5 (4.5)	117	113.7 (4.8)	116.1	115
7	123.8 (5.6)	126.6 (5.7)	119.6	—	128.7 (5.3)	126.7 (6.6)	123.5 (6.0)	123	120.5 (4.7)	121.7	122
8	128.9 (6.1)	130.3 (4.8)	127.2	—	134.0 (5.0)	130.3 (6.6)	132.7 (5.7)	128	125.3 (5.8)	127.0	128
9	134.7 (5.6)	135.7 (4.6)	131.1	—	141.8 (5.5)	136.3 (6.4)	138.0 (7.9)	133	130.0 (5.8)	132.2	133
10	140.2 (6.2)	140.3 (5.4)	137.7	—	146.0 (6.8)	139.7 (6.2)	142.8 (6.7)	139	135.1 (6.3)	137.5	139
11	146.2 (7.9)	145.0 (7.6)	144.0	144.4 (6.5)	150.9 (7.0)	—	145.0 (6.1)	143	141.9 (5.3)	143.3	144
12	155.0 (8.7)	153.1 (8.1)	145.9	148.8 (6.2)	156.3 (8.0)	—	153.4 (7.4)	149	146.8 (7.1)	149.7	149

TABLE 8.3 (continued)
Average Stature in cm (With Standard Deviation) of Male Children and Adolescents)

Age (years)	C	F	G	J	N	P	S	UK	US1	US2	US3
13	161.2	158.4	153.3	150.0	—	—	158.2	155	149.5	156.5	156
	(7.5)	(9.0)		(6.9)			(9.0)		(7.8)		
14	166.5	164.5	161.5	164.1	—	—	170.4	163	—	163.1	164
	(6.4)	(8.9)		(6.8)			(8.2)				
15	169.0	168.6	166.5	167.5	—	—	173.5	169	—	169.0	170
	(6.2)	(6.9)		(5.6)			(5.4)				
16	171.0	173.0	171.5	167.5	—	—	179.9	173	—	173.5	173
	(5.8)	(7.1)		(6.0)			(5.5)				
17	171.7	174.3	173.6	169.7	—	—	—	175	—	176.2	175
	(5.3)	(5.7)		(4.0)							
18	—	175.9	175.8	170.8	—	—	—	176	–	176.8	176
		(7.6)		(5.7)							

The curves in Figures 8.2 through 8.5 rely on data collected since the 1970s, thoroughly updated in recent years (Kuczmarski et al., 2002). These charts illustrate the initial fast growth of infants and young children and the following slowing of that development until adult body dimensions are attained. For 20-year-olds, the CDC charts show average statures of almost 177 cm for males (Figure 8.4) and 163+ cm for females (Figure 8.5). These numbers agree with those for U.S. Army soldiers: 176 cm for men and almost 163 cm for women (see Table A.10 in the Appendix). The soldiers were measured in 1988, and ongoing secular increases in average stature make the slightly larger numbers shown in the current CDC growth charts plausible.

The statistically experienced observer recognizes the large variation in children's dimensions from the wide spread of data on both sides of the average trend, a fact that is easily overlooked due to the apparent smoothness of the calculated growth curves.

BODY MASS OF CHILDREN AND ADOLESCENTS

Biomechanical information on changes in body mass with age, including the location of the center of mass of the body while standing or sitting, is of special importance in the design of products such as baby strollers and restraint devices in automobiles. Data concerning body mass have been compiled in Table 8.4.

Naturally, as children get older and taller, their body masses enlarge, and the absolute heights of the center of mass when they stand or sit also increase. However, Table 8.4 (excerpted from the 1975 data of Snyder et al.) shows three somewhat surprising facts:

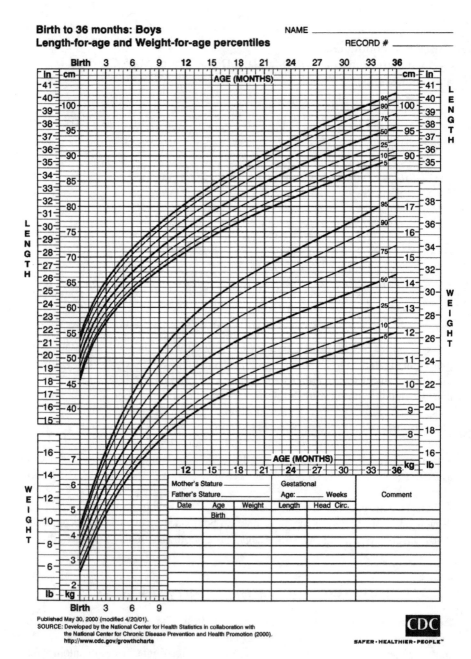

FIGURE 8.2 Growth chart for infant boys in the United States. (From "Clinical Growth Charts [U.S. children, birth to 20 years]," CDC [Centers for Disease Control and Prevention], 2001, Hyattsville, MD: National Center for Health Statistics. Retrieved from http://www.cdc.gov/nchs/about/major/nhanes/growthcharts/clinical_charts.htm.)

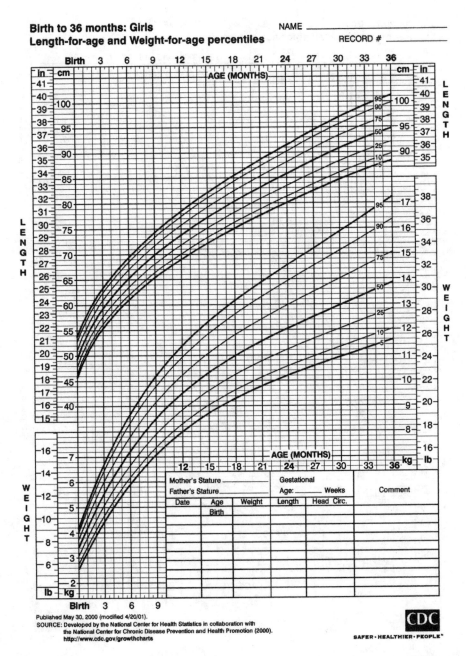

FIGURE 8.3 Growth chart for infant girls in the United States. (From "Clinical Growth Charts [U.S. children, birth to 20 years]," CDC [Centers for Disease Control and Prevention], 2001, Hyattsville, MD: National Center for Health Statistics. Retrieved from http://www.cdc.gov/nchs/about/major/nhanes/growthcharts/clinical_charts.htm.)

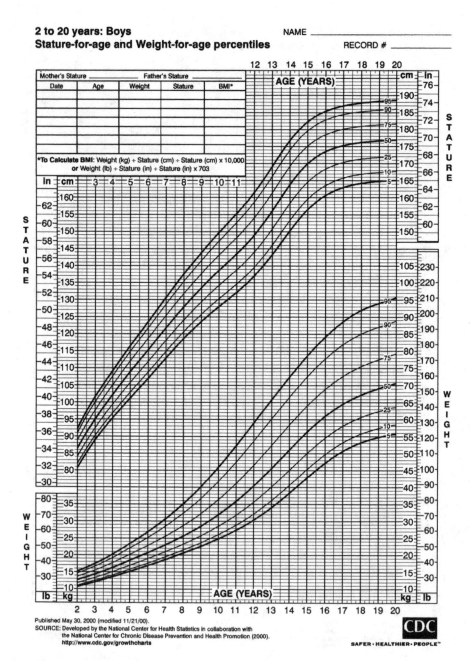

FIGURE 8.4 Growth chart for boys aged from 2 to 20 years in the United States. (From "Clinical Growth Charts [U.S. children, birth to 20 years]," CDC [Centers for Disease Control and Prevention], 2001, Hyattsville, MD: National Center for Health Statistics. Retrieved from http://www.cdc.gov/nchs/about/major/nhanes/growthcharts/clinical_charts.htm.)

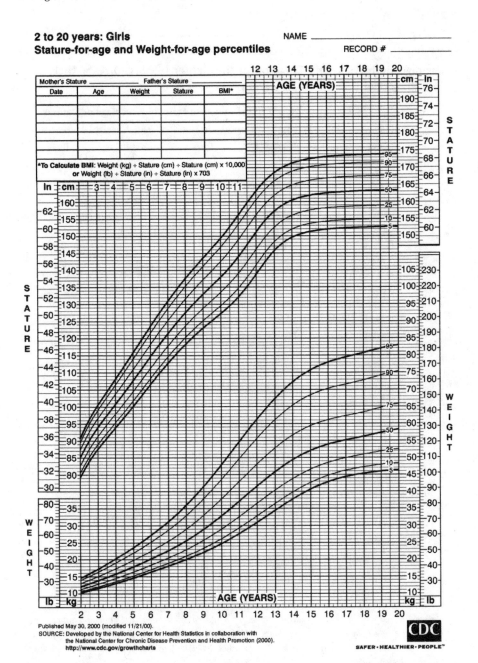

FIGURE 8.5 Growth chart for girls aged from 2 to 20 years in the United States. (From "Clinical Growth Charts [U.S. children, birth to 20 years]," CDC [Centers for Disease Control and Prevention], 2001, Hyattsville, MD: National Center for Health Statistics. Retrieved from http://www.cdc.gov/nchs/about/major/nhanes/growthcharts/clinical_charts.htm.)

TABLE 8.4
Average Body Mass and Location of the Center of Mass of U.S. Girls and Boys (With Standard Deviation)

			Height of the Center of Mass (in Percentage of Stature)			
Age	Body Mass (*kg*)		Standing (Height Above Floor)		Seated (Height Above Seat)	
(years)	Girls	Boys	Girls	Boys	Girls	Boys
Birth	4.6	4.8	59.4	58.5	50.2	48.0
	(1.1)	(1.2)	(1.9)	(2.1)	(3.3)	(4.4)
0.5	6.7	7.4	58.1	59.1	47.1	46.6
	(0.9)	(0.9)	(2.5)	(2.3)	(2.8)	(3.7)
1	8.9	9.5	58.1	58.5	44.6	45.6
	(1.3)	(0.8)	(1.8)	(2.4)	(2.3)	(2.7)
2	11.2	12.2	57.5	57.5	41.3	39.3
	(1.1)	(1.2)	(1.0)	(1.1)	(2.4)	(2.6)
3	12.8	14.2	59.3	58.9	39.1	37.6
	(1.1)	(1.5)	(2.5)	(1.0)	(1.9)	(1.2)
4	15.4	15.8	58.8	59.7	37.9	37.2
	(1.8)	(1.8)	(2.0)	(1.7)	(3.3)	(2.2)
5	17.7	18.3	59.3	58.9	35.3	36.6
	(2.3)	(2.1)	(2.0)	(1.7)	(2.6)	(2.6)
6	19.3	20.8	59.3	59.1	34.0	35.0
	(2.7)	(3.0)	(1.5)	(1.3)	(2.3)	(1.9)
7	21.8	23.2	58.6	58.7	33.3	33.1
	(2.7)	(3.1)	(1.1)	(1.3)	(1.8)	(2.1)
8	24.2	25.3	58.0	58.6	32.2	32.3
	(4.0)	(4.4)	(1.7)	(1.1)	(2.4)	(2.2)
9	27.7	27.7	58.0	57.9	30.7	32.1
	(5.2)	(4.6)	(1.4)	(1.1)	(1.6)	(1.8)
10	30.6	30.4	57.5	58.0	30.2	31.1
	(5.8)	(5.2)	(0.9)	(1.1)	(2.0)	(1.9)
11	34.4	35.4	57.4	57.7	29.5	30.0
	(7.2)	(5.8)	(0.7)	(1.0)	(1.6)	(1.6)
12	38.1	38.8	57.4	57.8	29.4	30.1
	(7.3)	(6.4)	(1.1)	(1.0)	(1.4)	(2.1)
13	48.0	40.7	57.4	58.0	29.2	29.7
	(8.1)	(7.0)	(1.3)	(1.5)	(1.3)	(1.5)

Data excerpted from *Physical Characteristics of Children as Related to Death and Injury for Consumer Product Safety Design* (Final Report, UM-HSRI-BI-75-5), by R. G. Snyder, M. L. Spencer, C. L. Owings, and L. W. Schneider, 1975, Ann Arbor, MI: The University of Michigan, Highway Safety Research Institute.

- For the standing body, the average height of the center of mass remains, from birth to 13 years of age, consistently at between 57% and 60% of stature.
- For the sitting body, the average height of the center of mass diminishes from near 50% of stature at birth to about 30% of stature at the ages of 10 or 11 years and then remains unchanged until the age of 13 years.
- All mass-related developments are essentially the same for young girls and boys. This finding suggests that the designer of body restraint systems does not need to differentiate between boys and girls until the onset of puberty with its physical changes.

BODY STRENGTH OF CHILDREN
AND ADOLESCENTS

Infants' movements are fairly uncoordinated, and they do not possess defined body strengths that allow systematic measurements, but their energetic and unexpected movements can inflict damage on others and themselves. Controlled body strength develops quickly during early and middle childhood. Peebles and Norris (1998, 2000, 2003) and researchers at the University of Nottingham (2002) measured the maximal exertions of British children and adolescents (as well as of adults up to 86 years) in six static strength tests. With altogether only about 150 participants, the number of participants in each age bracket was small, but the results plausibly illustrate the increase of muscle strength expected from 2 years to about 15 years of age.

Table 8.5 lists measurements of handgrip and grasp forces measured by Owings, Chaffin, Snyder, and Norcutt (1975) in the early 1970s on several hundred U.S. children. Nearly a quarter century later, Haeger-Ross and Roesblad (2002) reported similar grip strengths exerted by Swedish boys and girls, aged from 4 to 16 years, with about 20 participants in each one-year age group. Table 8.6 compiles these data for a comparison.

Table 8.7 reports torques about body joints that several hundred U.S. children of age 3 to 10 years exerted during tests conducted in the early 1970s (Owings et al., 1975).

For both genders, Tables 8.5 through 8.7 indicate a strong positive correlation between age and static muscle strength. However, statistical tests show no significant differences between younger boys' and girls' exertions: Unisex strength tables appear appropriate up to the age of about 10 years. Likewise, right or left hand preferences for handwriting and ball throwing are practically the same until that age in both genders. However, the large standard deviations in the three strength tables reflect the great interindividual variations that exist among the children.

Beginning at 10 to 12 years of age, on average, boys become discernibly stronger than girls, yet most strengths are not significantly correlated with each other, meaning that a person's score in one test does not well predict an exertion under other conditions or done with different body segments (Ager, Olivett, & Johnson, 1984; O. J. Bowman & Katz, 1984; Burke, Tuttle, Thompson, Janney, & Weber, 1953;

TABLE 8.5
Average Side Grip and Grasp Forces in N (With Standard Deviations) Exerted by U.S. Children, Boys and Girls Combined (*n* = 227)

Age (years)	Thumb–Forefinger Side Grip ("Side Pinch"*) N	Power Grasp ("Grip Strength") N
3	18.6	45.1
	(4.9)	(14.7)
4	26.5	57.9
	(5.9)	(17.7)
5	31.4	71.9
	(7.8)	(18.6)
6	38.3	89.3
	(5.9)	(22.6)
7	41.2	105.0
	(6.9)	(32.4)
8	47.1	124.6
	(9.8)	(33.4)
9	52.0	145.2
	(9.8)	(35.3)
10	51.0	163.8
	(8.8)	(37.3)

* Pinch surfaces 20 mm apart.

Data excerpted from *Strength Characteristics of U.S. Children for Product Safety Design* (Report 011903-F), by C.L. Owings, D. B. Chaffin, R. G. Snyder, and R. Norcutt, 1975, Ann Arbor, MI: The University of Michigan.

Daams, 2001; Haeger-Ross and Roesblad, 2002; Imrhan, 1986; Lowrey, 1986; Malina & Roche, 1983; Norris & Wilson, 1995; Owings et al., 1975; Peebles & Norris, 2000, 2003; Proos, 1993; Roche & Malina, 1983; Snyder et al., 1975).

DESIGNING FOR CHILDREN AND ADOLESCENTS

Proper use of ergonomic data about children and adolescents in the design process is important for ensuring that the final products suit their users. Information on user characteristics and abilities such as body sizes, muscular strengths, and motor skills is needed to ensure that products are safe. For example, small children should not be able to break parts off toys and then possibly swallow them. Unfortunately, until the 1970s, little ergonomic information about children was available that had been systematically and comprehensively measured. Then the U.S. Consumer Product Safety Commission sponsored a series of studies (Owings et al., 1975; Schneider Lehman, Pflug, & Owings, 1986; Snyder et al., 1975, 1977) that provided essential

TABLE 8.6
Average Power Grasp Force in N (With Standard Deviations) of Swedish Girls and Boys ("Peak Grip Strength")

Age (years)	Girls' Strength	Boys' Strength
4	64.5 (8.9) n = 10	57.9 (15.6) n = 11
5	64.0 (13.0) n = 23	69.0 (17.1) n = 20
6	82.0 (22.7) n = 22	83.1 (18.3) n = 15
7	91.8 (14.1) n = 24	103.1 (20.4) n = 21
8	106.2 (21.6) n = 17	125.5 (21.8) n = 18
9	127.4 (16.1) n = 20	136.9 (26.0) n = 27
10	121.0 (28.9) n = 20	167.1 (27.3) n = 25
11	165.7 (39.8) n = 21	187.6 (39.5) n = 20
12	177.3 (46.3) n = 23	219.9 (38.1) n = 20
13	233.0 (62.9) n = 22	274.9 (65.7) n = 20
14	282.5 (54.0) n = 24	343.4 (74.7) n = 27
15	288.9 (34.3) n = 21	414.7 (84.1) n = 21
16	322.0 (53.7) n = 16	490.6 (74.9) n = 22

Adapted from "Norms for Grip Strength in Children Aged 4–16 Years," by C. Haeger-Ross and B. Roesblad, 2002, *Acta Paediatrica, 91*, 617–625.

information on children. Yet, for the last three decades, no new survey data on American children have been published, and surprisingly little is systematically known about adolescents. In Europe, Steenbekkers (1993) addressed child development and its design implications, especially for accident prevention. In the United Kingdom, the Department of Trade and Industry (DTI) sponsored a collection of British and international data and published these in handbooks on children and adults (Norris & Wilson, 1995; Peebles & Norris, 1998, 2000, 2003; S. Smith, Norris, & Peebles, 2000; University of Nottingham, 2002).

DESIGNING SAFE OPENINGS

Head, neck, and hand entrapments injure many children. To generate data for safe openings, the U.S. Consumer Product Safety Commission sponsored an anthropometric study of American children. The 1986 report of Schneider et al. describes the sampling and measuring strategies employed and provides the body dimensions gathered. Head breadth, chest depth, and hand diameter are the critical dimensions that determine the design of openings, such as the distance between the stakes of railings that do not allow children to pass through them.

Measurements of head breadth, chest depth, and hand clearance diameter of American children appear in Table 8.8. That listing shows that up to the age of 12 years, girls have narrower heads, shallower chests, and smaller hand clearance diameters

TABLE 8.7

Average Torques in *Ncm* (With Standard Deviations) Around Wrist, Elbow, and Knee Exerted by U.S. Children, Girls and Boys Combined

Age (years)	Wrist		Elbow		Knee	
	Flexion	Extension	Flexion	Extension	Flexion	Extension
3	84	63	606	616	500	1673
	(47)	(22)	(156)	(111)	(197)	(616)
4	122	61	731	724	468	1866
	(61)	(28)	(233)	(259)	(194)	(710)
5	152	69	932	901	706	2301
	(79)	(30)	(319)	(285)	(351)	(738)
6	224	90	1192	1034	956	2717
	(85)	(40)	(299)	(373)	(386)	(961)
7	268	113	1687	1332	1175	3788
	(105)	(47)	(415)	(441)	(334)	(1165)
8	352	122	2114	1612	1371	4762
	(128)	(44)	(506)	(437)	(564)	(1391)
9	453	167	2248	1676	1986	5648
	(188)	(74)	(674)	(527)	(638)	(1386)
10	434	164	2362	1596	2084	5553
	(166)	(41)	(603)	(446)	(842)	(1826)
	n = 211	*n* = 205	*n* = 495	*n* = 496	*n* = 267	*n* = 496

Data excerpted from *Strength Characteristics of U.S. Children for Product Safety Design* (Report 011903-F), by C. L. Owings, D. B. Chaffin, R. G. Snyder, and R. Norcutt, 1975, Ann Arbor, MI: The University of Michigan.

than do boys. If openings are narrow enough not to let small girls' bodies pass, boys should certainly not be able to squeeze through either. Note that this is a good example of designing for a single value, here a minimum, which was discussed in Chapter 3.

DESIGNING FURNITURE FOR PLAY AND WORK

Designing furniture suitable for children became of strong interest to physicians more than a century ago as a side effect of medical evaluations of adult farmers and laborers, many of whom were found to suffer from deformed spines. Orthopedists saw various forms of scolioses, flattened spinal columns or exaggerated lordoses and kyphoses, and other symptoms of spinal damage and disease. The reasons were traced to malnutrition and harmful postures at work, often bent and twisted stances maintained over extended periods (Bonne, 1969). From this the concern arose that children might also grow up with deformed spinal columns when they sat for long periods, such as in the classroom, in "collapsed, crouched, slumped, skewed, stooped, round-shouldered, and flat-chested carriage." Instead, the child should sit completely

TABLE 8.8
Average Head Breadth, Chest Depth, and Hand Clearance Diameter in cm (With Standard Deviations) of U.S. Children, Girls and Boys Combined

Age (years)	Head Breadth		Chest Depth		Hand Clearance Diameter	
	Girls	Boys	Girls	Boys	Girls	Boys
Birth	10.3	10.4	9.0	9.3	3.21	3.33
	(0.6)	(0.7)	(0.9)	(0.9)	(0.29)	(0.30)
0.5	11.4	11.7	9.9	9.9	3.55	3.72
	(0.6)	(0.6)	(0.9)	(0.9)	(0.28)	(0.26)
1	12.3	12.6	10.4	11.0	3.86	4.14
	(0.4)	(0.6)	(1.1)	(0.6)	(0.25)	(0.29)
2	13.0	13.3	11.3	11.6	4.10	4.24
	(0.5)	(0.4)	(1.0)	(1.0)	(0.27)	(0.32)
3	13.3	13.5	11.8	12.0	4.30	4.51
	(0.5)	(0.4)	(0.8)	(1.2)	(0.32)	(0.24)
4	13.5	13.8	12.2	12.5	4.50	4.57
	(0.4)	(0.4)	(0.8)	(0.9)	(0.27)	(0.28)
5	13.6	14.0	12.7	13.0	4.66	4.82
	(0.4)	(0.5)	(1.1)	(1.0)	(0.30)	(0.32)
6	13.7	14.0	13.2	13.3	4.79	4.99
	(0.4)	(0.4)	(1.0)	(1.1)	(0.28)	(0.30)
7	13.9	14.2	13.5	14.1	5.01	5.16
	(0.4)	(0.5)	(1.0)	(1.1)	(0.31)	(0.30)
8	14.0	14.2	13.7	14.3	5.08	5.28
	(0.4)	(0.5)	(1.4)	(1.3)	(0.34)	(0.38)
9	14.1	14.3	14.4	14.8	5.22	5.42
	(0.5)	(0.4)	(1.4)	(1.3)	(0.33)	(0.35)
10	14.1	14.4	14.7	15.2	5.42	5.56
	(0.5)	(0.5)	(1.5)	(1.3)	(0.33)	(0.33)
11	14.2	14.6	15.7	16.2	5.60	5.85
	(0.4)	(0.4)	(2.0)	(1.6)	(0.40)	(0.38)
12	14.5	14.5	16.2	16.8	5.82	6.03
	(0.6)	(0.5)	(1.7)	(1.6)	(0.34)	(0.37)
13	14.6	14.5	17.9	17.2	6.16	6.06
	(0.5)	(0.4)	(2.2)	(1.7)	(0.37)	(0.40)

Data excerpted from *Physical Characteristics of Children as Related to Death and Injury for Consumer Product Safety Design* (Final Report, UM-HSRI-BI-75-5), by R. G. Snyder, M. L. Spencer, C. L. Owings, and L.W. Schneider, 1975, Ann Arbor, MI: The University of Michigan, Highway Safety Research Institute.

upright, which was commonly called hygienic and considered healthy, as Zacharkow (1988) reported in detail.

When looking at an erect standing healthy human from the left side, the normal spine appears bent somewhat like the letter S, but when viewed from the front or rear, the spine is straight. In spite of the normal curvature, apparent in the side view, the ideal of the erect back as a sign of healthy standing and sitting positions became commonly accepted, medically as well as socially, for children and adults in the 1880s (more about this in K. H. E. Kroemer & A. D. Kroemer, 2001a; K. H. E. Kroemer, H. B. Kroemer, & Kroemer-Elbert, 2001). Since then, admonitions to children and teenagers to sit up, sit upright, stand erect, and straighten up have been common slogans. Accordingly, furniture was designed in the late 19th century and throughout most of the 20th century to promote, and even impose, the upright posture (Zacharkow, 1988).

Figure 8.6 shows a combined seat and writing desk recommended in 1888 by Lorenz for adolescents. The desk surface is elevated to about chest height and angled to facilitate reading (with erect head and neck) and to support the forearms when writing. A tall backrest, curved to follow the natural lumbar lordosis, supports the trunk. The seat surface reclines slightly and has a shape that follows the form of the buttocks and thighs. The thighs are nearly horizontal and the lower legs about vertical.

For most of the last century, teachers and designers of school and office furniture believed in an idol of "correct sitting": an erect stance of the upper body and head, and 90° in hip, knee, and ankle joints (see the discussion in Chapter 4). This stilted pose was considered appropriate for both children and adults. The basic chair model, designed to promote (even enforce) this stiff posture, remained essentially the same into the early computer era. The inclined reading/writing surface of the late 1800s was lowered and made horizontal to accommodate typewriters and then computers (K. H. E. Kroemer & A. D. Kroemer, 2001a).

School Furniture

Children's postures, especially when seated in school, have been a concern since the 18th century, as Zacharkow (1988) described. His book contains many illustrations of 19th- and early-20th-century furniture designed to make children assume "hygienic" postures when sitting, half-sitting, or standing. Figure 8.7 is an example of the sophisticated designs of that period. An upright posture, especially a straight trunk and neck, was the accepted correct pose.

However, as early as 1888, Lorenz wrote, to achieve "healthful, comfortable, and graceful" sitting, "the upright (military) sitting position ... is too rigorous, calls for an excess amount of muscular exertion, and does not afford sufficient support to the back of the child.... The reclined-sitting position, in which the back is supported at all times by a properly curved back-support inclined backwards from the inclined seat-surface at an angle of 10 to 15 degrees is to be recommended" (Zacharkow, 1988, p. 148). It is surprising that *sitting upright* over extended periods remained the guiding principle for designers of furniture, in schools and offices, for almost a full century in spite of the apparent discomfort associated with this posture.

Finally, in recent years, new design axioms regarding chairs have been widely accepted (HFES, 2002; K. H. E. Kroemer & A. D. Kroemer, 2001a, 2001b; Kroemer et al., 2001). Chairs should

FIGURE 8.6 Seat–desk combination proposed in 1888 by Lorenz. (Courtesy of H. W. Juergens)

- encourage free-flowing motions instead of imposing a single static posture,
- accommodate individual preferences for comfort, and
- support variability among users.

Body dimensions of children are important for the design of furniture, especially for use in schools (Steenbekkers & Molenbroek, 1990). Commonly, children of very different body sizes attend the same class. Thus, if adjustable furniture is not available, tables and chairs of various sizes should provide individual accommodation. In reality, furniture of different sizes is rarely available to the children. Even if available, its proper use poses problems for a variety of organizational and other practical reasons. For example, young children might find it difficult to select or

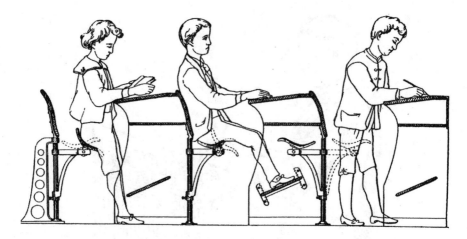

FIGURE 8.7 School furniture proposed in 1890 by Schindler. Note the adjustment features of seat and footrest. (Courtesy of H. W. Juergens)

adjust furniture to their size and liking; adult expertise, help, and supervision are essential.

In spite of these practical issues, the efforts to design and use suitable chairs and seats remain vigorous, fueled by the hope that "good" furniture will promote proper use by schoolchildren and, in turn, advance good health, and even enhance their performance, as typified by recent publications (e.g., De Looze, Kuijt & Van Dieen-Evers, 2003; Helander, 2003; Helander & Tham, 2003; Knight & Noyes, 1999; Milanese & Grimmer, 2004; Moelenbroek, Kroon-Ramaekers, & Snijders, 2003; Troussier, Tesniere, Fauconnier, Grison, Juvin, & Phelip, 1999) and the efforts of special committees of the International Ergonomics Association and the Human Factors and Ergonomics Society.

Computer Play Stations and Workplaces

In the United States, similar to other countries, many children and teenagers use computers. The computer is now an important toy at home, a major communication device, and an indispensable instructional and learning tool. Unfortunately, the fundamental problems of the standard keyboard, inherited from the 1878 QWERTY design, were not remedied before its use became widespread (Kroemer, 2001).

It is of some concern that small children often sit at workstations that were originally designed for adults, with seats and tables that do not fit their bodies because the furniture is too big (Hedge, Barrero, & Maxwell, 2000; K. H. E. Kroemer & A. D. Kroemer, 2001a; Saito, Sotoyama, Jonai, Akutsu, Yatani, & Marumoto, 2000; Straker, Harris, & Zandvliet, 2000). Figure 8.8 illustrates this.

The mistaken 1990s practice of putting the monitor on top of the CPU (central processing unit) exacerbates the situation. Children must crane their necks to look up at the display even if the seat is in its highest position, which usually makes it

FIGURE 8.8 Child at a computer workstation. In spite of the attempt to raise the body by kneeling on the seat, the child cannot lean against the backrest and must elevate the eyes to observe the monitor. This brings about an excessive backward bend (lordosis) of the spinal column in the neck region. Furthermore, note the pressure on the arms at the edges of the table and the bent wrist as the child holds the computer mouse.

too tall for their feet to reach the floor. Furthermore, the regular computer keyboard, with its large size (to accommodate more than 100 keys), does not fit the child's small hands and short reach. Instead of using a desktop computer, a laptop might better suit the child, with regard to both the size of the keyboard and the location of the screen. Computer workstations designed for children's bodies and that adjust to their relatively quickly changing body dimensions are not yet often available in furniture stores but can be found in catalogs and through the Internet.

The International Ergonomics Association (http://iea.cc) has a technical committee that focuses on designs suitable for children and adolescents. Other IEA standing committees are concerned with aging, rehabilitation, hospitals, safety, and other subjects that are themes in this book. The IEA can provide the addresses of federated national ergonomics societies in all regions of the globe, and of related journals and conferences. In the United States, the Human Factors and Ergonomics Society (http://hfes.org) has similar committees, as do many other national societies. To all of these endeavors, the HFES motto applies: "People-friendly design through science and engineering."

SUMMARY

Ergonomically useful, systematic, and comprehensive hard data on children exist (even though most are a bit dated), but, surprisingly, not much information on teenagers is available. During childhood and adolescence, interindividual variability is wide and intraindividual changes occur quickly. Thus, it is not meaningful to compile ergonomics information on prepubescent children by age or gender, even though this is often done. Classifications of children and adolescents according to size, strength, or other performance criteria would be more useful.

Chapter 8 contains information on available anthropometric and biomechanical descriptors of children that are useful to design products such as railings, strollers, carrying seats, or helmets that must fit and protect the child. Up to the onset of puberty, these data are generally applicable to both boys and girls. The information also helps to design, select, and adjust furniture for computer play and workstations used at home and in school.

References

Ackerman, D. (1990). *A natural history of the senses*. New York: Random House.

Adams, M. A., Bogduk, N., Burton, K., & Dolan, P. (2002). *The biomechanics of back pain*. Edinburgh: Churchill Livingstone.

Ager, C. L., Olivett, B. L., & Johnson, C. L. (1984). Grasp and pinch strength in children 5 to 12 years old. *American Journal of Occupational Therapy, 38,* 107–113.

Al-Haboubi, M. (1991). Anthropometric study for the user population in Saudi Arabia. In *Proceedings of the 11th Congress of the International Ergonomics Association* (pp. 891–893). London: Taylor & Francis.

Al-Haboubi, M. H. (1999). Statistics for a composite distribution in anthropometric studies: The general case. *Ergonomics 42,* 565–572.

Amditis, A., Bekiaris, E., Braedel, C., & Knauth, P. (2003). Dealing with the problems of elderly workforce — The RESPECT approach. In H. Strasser, K. Kluth, H. Rausch, & H. Bubb (Eds.), *Quality of work and products in enterprises of the future* (pp. 881–884). Stuttgart, Germany: Ergonomia Verlag.

Andersen, T. B., Schibye, B., & Skotte, J. (2001). Sudden movements of the spinal column during health-care work. *International Journal of Industrial Ergonomics, 38,* 47–53.

Annis, J. F. (1996). Aging effects on anthropometric dimensions important to workplace design. *International Journal of Industrial Ergonomics, 18,* 381–388.

Annis, J. F., Case, H. W., Clauser, C. E., & Bradtmiller, B. (1991). Anthropometry of an aging workforce. *Experimental Aging Research, 17,* 157–176.

Annis, J. F., & McConville, J. T. (1996). Anthropometry. In A. Bhattacharya & J. D. McGlothlin (Eds.), *Occupational ergonomics* (chap. 1, pp. 1–46). New York: Marcel Dekker.

Armstrong, T. J. (2000). Analysis and design of jobs for control of work related musculoskeletal disorders (WMSDs). In F. Violante, T. Armstrong, & A. Kilbom (Eds.), *Occupational ergonomics: Work related musculoskeletal disorders of the upper limb and back* (chap. 5, pp. 51–81). London: Taylor & Francis.

Arndt, B., & Putz-Anderson, V. (2001). *Cumulative trauma disorders* (2nd ed.). London: Taylor & Francis.

Arswell, C. M., & Stephens, E. C. (2001). Information processing. In W. Karwowski (Ed.), *International encyclopedia of ergonomics and human factors* (pp. 256–259). London: Taylor & Francis.

Astrand, P. O., & Rodahl, L. (1986). *Textbook of work physiology* (3rd ed.). New York: McGraw-Hill.

August, S., & Weiss, P. L. (1992). A human factors approach to adapted access device prescription and customization. *Journal of Rehabilitation Research and Development, 29,* 64–77.

Backs, R. W., & Boucsein, W. (Eds.) (2003a). *Engineering psychophysiology: Issues and applications*. Mahwah, NJ: Erlbaum.

Backs, R. W., Lenneman, J. K., Wetzel, J. M., & Green, P. (2003b). Cardiac measures of driver workload during simulated driving with and without visual occlusion. *Human Factors, 45,* 525–538.

Bailey, R. W. (1996). *Human performance engineering*. Upper Saddle River, NJ: Prentice Hall.

Bakker, R. (1997). *Elderdesign: Designing and furnishing a home for your later years*. New York: Viking Penguin.

Ballard, B. (1995). How odor affects performance: A review. In *Proceedings of the ErgoCon '95, Silicon Valley Ergonomics Conference and Exposition* (pp. 191–200). San Jose, CA: San Jose State University.

Barlow, A. M., & Braid, S. J. (1990). Foot problems in the elderly. *Clinical Rehabilitation, 4*, 217–222.

Barnes, M. E., & Wells, W. (1994, April). If hearing aids work, why don't people use them? *Ergonomics in Design*, 18–24.

Barzun, J. (2000). *From dawn to decadence: 500 years of Western cultural life. 1500 to the present*. New York: HarperCollins.

Batavia, A. I., & Hammer, G. S. (1990). Toward the development of consumer-based criteria for the evaluation of assistive devices. *Journal of Rehabilitation Research and Development, 27*, 425–436.

Baucom, A. H. (1996). *Hospitality design for the graying generation*. New York: Wiley.

Bazar, A. R. (1978). *Ergonomics in rehabilitation engineering* (Special Publication of Tech Briefs. Rehabilitation Engineering Center, Cerebral Palsy Research Foundation of Kansas). Wichita: Wichita State University.

Belsky, J. K. (1990). *The psychology of aging: Theory, research, and interventions*. Pacific Grove, CA: Brooks/Cole.

Berger, E. H., Royster, L. H., Royster, J. D., Driscoll, D. P., & Layne, M. (2003). *The noise manual* (5th ed.). Fairfax, VA: American Industrial Hygiene Association.

Bhattacharya, A., & McGlothlin, J. D. (Eds.). (1996). *Occupational ergonomics — Theory and applications*. New York: Marcel Dekker.

Birren, J. E. (Ed.). (1996). *Encyclopedia of gerontology*. San Diego, CA: Academic.

Birren, J. E., & Schaie, K. W. (Eds.). (2001). *Handbook of the psychology of aging* (5th ed.). San Diego, CA: Academic.

Blaikie, A. (1993). Images of age. A reflective process. *Applied Ergonomics, 24*, 51–57.

Bloswick, D. S., Shirley, B., & King, E. (1998). Medical and rehabilitation equipment case study: Design, development, and usability testing of a lift-seat wheelchair. In V. J. B. Rice (Ed.), *Ergonomics in health care and rehabilitation* (chap. 14, pp. 249–263). Boston: Butterworth-Heinemann.

Boff, K. R., Kaufman, L., & Thomas, J. P. (Eds.). (1986). *Handbook of perception and human performance*. New York: Wiley.

Boff, K. R., & Lincoln, J. E. (1988). *Engineering data compendium: Human perception and performance*. Wright-Patterson Air Force Base, OH: Armstrong Aerospace Medical Research Laboratory.

Bonne, A. J. (1969). On the shape of the human vertebral column. *Acta Orthopaedica Belgica, 35, Fasc. 3–4*, 567–583.

Borg, G. (2001). Rating scales for perceived physical effort and exertion. In W. Karwowski (Ed.), *International encyclopedia of ergonomics and human factors* (pp. 538–541). London: Taylor & Francis.

Bowman, O. J., & Katz, B. (1984). Hand strength and prone extension in right-dominant, 6 to 9 years old. *American Journal of Occupational Therapy, 38*, 367–376.

Bowman, S. A., Gortmaker, S. L., Ebbeling, C. B., Pereira, M. A., & Ludwig, D. S. (2004). Effects of fast food consumption on energy intake and diet quality among children in a national household survey. *Pediatrics, 113*, 112–118.

Bradtmiller, B. (2000). *Anthropometry for persons with disabilities: Needs in the twenty-first century*. Paper presented at RESNA 2000 Annual Conference and Research Symposium, Orlando, FL.

Bridger, R. S. (1995). *Introduction to ergonomics*. New York: McGraw-Hill.

Brody, J. (2004, January 27). Unhealthy trends result in the widening of America. *The Roanoke Times*, p. E 3.

Brown, J. R. (1972). *Manual lifting and related fields: An annotated bibliography*. Toronto: Labour Safety Council of Ontario, Ontario Ministry of Labour.

Brown, J. R. (1975). Factors contributing to the development of low-back pain in industrial workers. *American Industrial Hygiene Association Journal, 36*, 26–31.

Brownell, K. D. (2004). Fast food and obesity in children. *Pediatrics, 113,* 132.

Brummett, W. J. (1996). *The essence of home: Design solutions for assisted-living housing*. New York: Wiley.

Burke, W. E., Tuttle, W. W., Thompson, C. W., Janney, C. D., & Weber, R. J. (1953). The relation of grip strength and grip-strength endurance to age. *Applied Physiology, 5*, 628–630.

Burnsides, D. B., Boehmer, M., & Robinette, K. M. (2001). 3-D landmark detection and identification in the CAESAR project. In *Proceedings of the Third Internatinal Conference on 3-D Digital Imaging and Modeling* (pp. 393–398). Los Alamitos, CA: IEEE Computer Society.

Burry, H. C., & Stoke, J. C. J. (1985). Repetitive strain injury. *New Zealand Medical Journal, 98,* 601–602.

Byrne, M. D., & Gray, W. D. (2003). Returning human factors to an engineering discipline: Expanding the science base through a new generation of quantitative methods — preface to the special section. *Human Factors, 45,* 1–4.

Caccioppo, J. T., Tassinary L. G., & Berntson, G. G. (Eds.). (2000). *Handbook of psychophysiology* (2nd ed.). New York: Cambridge University Press.

Cai, D., & You, M. (1998). An ergonomic approach to public squatting-type toilet design. *Applied Ergonomics, 29*, 147–153.

Calhoun, G. L. (2001). Gaze-based control. In W. Karwowski (Ed.), *International encyclopedia of ergonomics and human factors* (pp. 234–236). London: Taylor & Francis.

Carayon, P., Haims, M. C., & Smith, M. J. (1999), Work organization, job stress, and work-related musculoskeletal disorders. *Human Factors, 41,* 644–663.

Casali, S. P., & Williges, R. C. (1990). Databases of accommodative aids for computer users with disabilities. *Human Factors, 32,* 407–422.

Centers for Disease Control and Prevention (CDC). (2001). *Clinical growth charts [U.S. children, birth to 20 years]* Hyattsville, MD: National Center for Health Statistics. (For the latest version, go to http://www.cdc.gov/nchs/about/major/nhanes/growth-charts/clinical_charts.htm)

Chaffin, D. B., Andersson, G. B. J., & Martin, B. J. (1999). *Occupational biomechanics* (3rd ed.). New York: Wiley.

Chakarbarti, D. (1997). *Indian anthropometric dimensions for ergonomic design practice*. Paldi, Ahmedabad, India: National Institute of Design. (Reviewed 1999 by J. Roebuck in *Ergonomics in Design,* April, p. 37. Data provided by J. Roebuck.)

Chan, A. H. S., So, R. S. Y., & Ng, E. N. S. (2000). Development and usability evaluation of an anthropometric database for Hong Kong Chinese. In *Proceedings of the 14th Triennial Congress of the International Ergonomics Association and 44th Annual Meeting of the Human Factors and Ergonomics Society* (pp. 6.287–6.290). Santa Monica, CA: Human Factors and Ergonomics Society.

Chapanis, A. (Ed.). (1975). *Ethnic variables in human factors engineering*. Baltimore: Johns Hopkins University Press.

Chapanis, A. (1996). *Human factors in systems engineering*. New York: Wiley.

Charlton, S. G., & O'Brien, T. G. (Eds.). (2002). *Handbook of human factors testing and evaluation*. Mahwah, NJ: Erlbaum.

Charney, W., & Hudson, M. A. (2004). *Back injury among health-care workers: Causes, solutions, and impacts.* Boca Raton, FL: CRC Press.

Chengular, S. N., Rodgers, S. H., & Bernard, T. E. (2003). *Kodak's ergonomic design for people at work* (2nd ed.). New York: Wiley.

Cherry, N. (1987). Physical demands of work and health complaints among women working late in pregnancy. *Ergonomics, 30,* 689–701.

Cheverud, J., Gordon, C. C., Walker, R. A., Jacquish, C., Kohn, L., Moore, A., & Yamashita, N. (1990). *1988 Anthropometric survey of U.S. Army personnel* (Technical Reports 90/031–036). Natick, MA: U.S. Army Natick Research, Development, and Engineering Center.

Clark, M. C., Czaja, S. J., & Weber, R. A. (1990). Older adults and daily living task profiles. *Human Factors, 32,* 537–549.

Clements, M. (1993, December). What we say about aging. *Parade Magazine,* 4–5.

Cole, T. J., Belizzi, M. C., Flegal, K. M., & Dietz, W. H. (2000, May 6). Establishing a standard definition of child overweight and obesity: International survey. *British Medical Journal, 320,* 1240–1243.

Coleman, R. (1993). A demographic overview of the ageing of first world populations. *Applied Ergonomics, 24,* 5–8.

Committee on an Aging Society. (Ed.). (1988). *America's aging — The social and built environment in an older society.* Washington, DC: National Academy Press.

Coniglio, I., Fubini, E., Masali, M., Masiero, C., Pierlorenzi, G., & Sagone, G. (1991). Anthropometric survey of Italian population for standardization in ergonomics. In *Proceedings of the 11th Congress of the International Ergonomics Association* (pp. 894–896). London: Taylor & Francis.

Craik, F. I. M., & Salthouse, T. A. (Eds.). (2000). *The handbook of aging and cognition* (2nd ed.). Mahwah, NJ: Erlbaum.

Cremer, R. (2001). Work design: Age-related policies. In W. Karwowski (Ed.), *International encyclopedia of ergonomics and human factors* (pp. 606–608). London: Taylor & Francis.

Culver, C. C., & Viano, D. C. (1990). Anthropometry of seated women during pregnancy. Defining a fetal region for crash protection research. *Human Factors, 32,* 625–636.

Czaja, S. J. (1990). *Human factors research for an aging population.* Washington, DC: National Research Council, National Academy Press.

Czaja, S. J. (1997). Using technology to aid in the performance of home tasks. In A. D. Fisk & W. A. Rogers (Eds.), *Handbook of human factors and the older adult* (chap. 13, pp. 311–334). San Diego, CA: Academic.

Czaja, S. J. (1998). Gerontology case study: Designing a computer-based communication system for older adults. In V. J. B. Rice (Ed.), *Clinical ergonomics* (chap. 9, pp. 143–134). Newton, MA: Butterworth-Heineman.

Czaja, S. J., (2001a). Telecommunication technology as an aid to family caregivers. In W. A. Rogers & A. Fisk (Eds.), *Human factors interventions for the health care of older adults* (chap. 9, pp.165–178). Mahwah, NJ: Erlbaum.

Czaja, S. J. (2001b). Ergonomics and older health care for older adults. In M. D. Mezey (Ed.), *The encyclopedia of care of elderly* (pp. 241–243). New York: Springer.

Czaja, S. J., & Lee, C. C. (2001). The Internet and older adults: Design challenges and opportunities. In N. Charness, D. C. Park, & B. A. Sabel (Eds.), *Aging and communication: Opportunities and challenges of technology* (pp. 60–81). New York: Springer.

Daams, B. J. (2001). Push and pull data and torque data. In W. Karwowski (Ed.), *International encyclopedia of ergonomics and human factors* (pp. 299–316, 334–342). London: Taylor & Francis.

Daltrov, L. H., Iversen, M. D., Larson, M. G., Lew, R., Wright, E., Ryan, J., Zwerling, C., Fossel, A. H., & Liang, M. H. (1997). A controlled trial of an educational program to prevent low back injuries. *New England Journal of Medicine, 337*, 322–328.

Daynard, D., Yassi, A., Cooper, J. E., Tate, R., Norman, R., & Wells, R. (2001). Biomechanical analysis of peak and cumulative spinal loads during simulated patient-handling activities: A substudy of a randomized controlled trial to prevent lift and transfer injury of health-care workers. *Applied Ergonomics, 32*, 199–214.

De Looze, M., Kuijt. L., & Van Dieen-Evers, J (2003). Sitting comfort and discomfort and the relationships with objective measures. *Ergonomics, 46*, 985–997.

Deyo, R. A., & Weinstein, J. N. (2001). Low back pain. *New England Journal of Medicine, 344*, 363–370.

DiNardi, S. R. (Ed.). (2003). *The occupational environment: Its evaluation, control, and management* (2nd ed.). Fairfax, VA: American Industrial Hygiene Association.

Driskell, J. E., & Radtke, P. H. (2003). The effect of gesture on speech production and comprehension. *Human Factors 45*, 445–454.

Dryden, R. D., & Kemmerling, P. T. (1990). Engineering assessment. In S. P. Sheer (Ed.), *Vocational assessment of impaired workers* (pp. 107–129). Aspen, CO: Aspen Press.

Eisenberg, A., Murkoff, H. E., & Hathaway, S. E. (1991). *What to expect when you are expecting* (2nd ed., pp. 205–206). New York: Workman.

Elford, W., Straker, L., & Strauss, G. (2000). Patient handling with and without slings: An analysis of the risk of injury to the lumbar spine. *Applied Ergonomics, 31,* 185–200.

ElKarim, M. A. A., Sukkar, M. Y., Collins, K. J., & Doré, C. (1981). The working capacity of rural, urban and service personnel in the Sudan. *Ergonomics, 24,* 945–952.

Elkind, J. I., Nickerson, R. S., Van Cott, H. P., & Williges, R. C. (1995). Employment and disabilities. In R. S. Nickerson (Ed.), *Emerging needs and opportunities for human factors research* (chap. 3, pp. 106–130). Washington, DC: National Academy Press.

Engkvist, I. L., Kjellberg, A., Wigaeus, H. E., Hagberg, M., Menckel, E., & Ekenvall, L. (2001). Back injuries among nursing personnel — Identification of work conditions with cluster analysis. *Safety Science, 37,* 1–18.

Errkola, R. (1976). The physical work capacity of the expectant mother and its effect on pregnancy, labour and the newborn. *International Journal of Obstetrics and Gynaecology, 14,* 153–159.

Evans, W. A. (1990). The relationship between isometric strength of Cantonese males and the US NIOSH guide for manual lifting. *Applied Ergonomics, 21,* 135–142.

Fast, A., Shapiro, M. D., & Edmond, J. (1987). Low back pain in pregnancy. *Spine, 12*, 368–371.

Faste, R. A. (1977). New system propels design for the handicapped. *Industrial Design,* 51–55.

Fathallah, F. A., & Brogmus, G. E. (1997). Hourly trends in occupational back pain and cumulative trauma disorders. In *Proceedings of the Human Factors and Ergonomics Society 41st Annual Meeting* (pp. 609–613). Santa Monica, CA: Human Factors and Ergonomics Society.

Feathers, D., Polzin, J., Paquet, V., Lenker, J., & Steinfeld, E. (2001). Comparison of traditional and electromechanical approaches for structural anthropometric data collection. In *Proceedings of the Human Factors and Ergonomics Society 45th Annual Meeting* (pp. 1036–1039). Santa Monica, CA: Human Factors and Ergonomics Society.

Feeney, R, Summerskill, S., Porter, M., & Freer. M. (2000). Designing for disabled people using a 3D human modelling CAD system. In K. Landau (Ed.), *Ergonomic software tools in product and workplace design. A review of recent developments in human modeling and other design aids* (pp. 195–203). Stuttgart, Germany: Ergon.

Fernandez, J. E., Malzahn, D. E., Eyada, O. K., & Kim, C. H. (1989). Anthropometry of Korean female industrial workers. *Ergonomics, 32,* 491–495.

Fernandez, J. E., & Uppugonduri, K. G. (1992). Anthropometry of South Indian industrial workmen. *Ergonomics, 35,* 1393–1398.

Fernie, G. (1997). Assistive devices. In A. D. Fisk & W. A. Rogers (Eds.), *Handbook of human factors and the older adult* (chap. 12, pp. 289–310). San Diego, CA: Academic.

Fisher, M. J. (2001, April 4). The ergonomic rocking chair. *The Atlantic Monthly, 297,* 93–95.

Fisk, A. D. (1999, January). Human factors and the older adult. *Ergonomics in Design,* 8–13.

Fisk, A. D., & Rogers, W. A. (Eds.). (1997). *Handbook of human factors and the older adult.* San Diego, CA: Academic.

Fisk, A. D., Rogers, W. A., Charness N., Czaja, S. J., & Sharit, J. (Eds.). (2004). *Designing for older adults.* Boca Raton, FL: CRC Press.

Flach, J. M., & Dominguez, C. O. (1995, July). Use-centered design. *Ergonomics in Design,* 19–24.

Fleishman, E. A., & Quaintance, M. K. (1984). *Taxonomies of human performance: The description of human tasks.* Orlando, FL: Academic.

Floru, R., & Cnockaert, (1991). *Introduction a la psychophysiologie du travail* [Introduction to the psychophysiology of work]. Nantes, France: Presses Universitaires de Nancy.

Fluegel, B., Greil, H., & Sommer, K. (1986). *Anthropologischer atlas.* Berlin, Germany: Tribuene.

Fontaine, K. R., Gadbury, G., Heymsfield, S. B., Kral, J., Albu, J. B., & Allison D. (2002). Quantitative prediction of body diameter in severely obese individuals. *Ergonomics, 45,* 49–60.

Foti, T., Davids, J. R., & Bagley, A. (2000). A biomechanical analysis of gait during pregnancy. *Journal of Bone Joint Surgery, 5,* 625–632.

Fozard, J. L., & Gordon-Salant, S. (2001). Changes in vision and hearing with aging. In J. E. Birren & K. W. Schaie (Eds.), *Handbook of the psychology of aging* (5th ed., chap. 10, pp. 241–266). San Diego, CA: Academic.

Franco, G., & Fusetti, L. (2004). Bernadino Ramazzini's early observations of the link between musculoskeletal disorders and ergonomic factors. *Applied Ergonomics, 35,* 67–70.

Freivalds, A. (1999). Ergonomics of hand controls. In W. Karwowski & W. S. Marras (Eds.), *The occupational ergonomics handbook* (chap. 27, pp. 461–478). Boca Raton, FL: CRC Press.

Froufe, T., Ferreira, F., & Rebelo, F. (2002). Collection of anthropometric data from primary schoolchildren. In *Proceedings of CybErg 2002, the 3rd International Cyberspace Conference on Ergonomics* (pp. 166–171). London: International Ergonomics Association Press and personal communications by F. Rebelo, February 2003.

Fry, H. J. H. (1986a). Overuse syndrome in musicians — 100 years ago: A historical review. *The Medical Journal of Australia, 145,* 620–625.

Fry, H. J. H. (1986b). Physical signs in the hands and wrists seen in the overuse injury syndrome of the upper limb. *Australian and New Zealand Journal of Surgery, 56,* 47–49.

Fry, H. J. H. (1986c). Overuse syndrome in musicians: Prevention and management. *The Lancet,* September 27, 723–731.

Fry, H. J. H. (1986d, March). What's in a name? The musician's anthology of misuse. *Medical Problems of Performing Artists,* 36–38.

Fry, H. J. H. (1986e, June). Incidence of overuse syndrome in the symphony orchestra. *Medical Problems of Performing Artists,* 51–55.

Gagnon, M., Akre, F., Chehade, A., Kemp, F., & Lortie, M. (1987). Mechanical work and energy transfers while turning patients in bed. *Ergonomics, 30,* 1515–1530.

Gallwey, T. J., & Fitzgibbon, M. J. (1991). Some anthropometric measures on an Irish population. *Applied Ergonomics, 22,* 9–12.

Gardner-Bonneau, D., & Gosbee, J. (1997). Health care and rehabilitation. In A. D. Fisk & W. A. Rogers (Eds.), *Handbook of human factors and the older adult* (chap. 10, pp. 231–255). San Diego, CA: Academic.

Garrett, J. W., & Kennedy, K. W. (1971). *A collation of anthropometry* (AMRL-TR-68-1). Wright-Patterson Air Force Base, OH: Aerospace Medical Research Laboratories.

Gawande, A. (1998, September 21). The pain perplex. *The New Yorker*, 86–94.

Gilad, I. (2001). Wheelchairs. In W. Karwowski (Ed.), *International encyclopedia of ergonomics and human factors* (pp. 981–984). London: Taylor & Francis.

Gite, L. P., & Yadav, B. G. (1989). Anthropometric survey for agricultural machinery design. *Applied Ergonomics, 20,* 191–196.

Gloss, D. A., & Wardle, M. G. (1982). Reliability and validity of American Medical Association's guide to ratings of permanent impairment. *Journal of the American Medical Association, 248,* 2292–2296.

Goebel, M., Fietz, A., & Friedorf, W. (2003). User adapted presentation of Internet information for older adults. In H. Strasser, K. Kluth, H. Rausch, & H. Bubb (Eds.), *Quality of work and products in enterprises of the future* (pp. 1067–1074). Stuttgart, Germany: Ergonomia Verlag.

Goenen, E., Kalinkara, V., & Oezgen, O. (1991). Anthropometry of Turkish women. *Applied Ergonomics, 22,* 409–411.

Gordon, C. C., Churchill, T., Clauser, C. E., Bradtmiller, B., McConville, J. T., Tebbetts, I., & Walker, R. A. (1989). *1988 Anthropometric survey of U.S. Army personnel: Summary statistics interim report* (Technical Report NATICK/TR-89/027). Natick, MA: U.S Army Natick Research, Development and Engineering Center.

Gordon, C. C., Corner, B. D., & Brantley, J. D. (1997). *Defining extreme sizes and shapes for body armor and load-bearing systems design: Multivariate analysis of U.S. Army torso dimensions* (Technical Report NATICK/TR-97/012). Natick, MA: U.S. Army Natick Research, Development and Engineering Center.

Green, R. J., Self, H. C., & Ellifritt, T. S. (1995). *50 Years of human engineering.* Wright-Patterson Air Force Base, OH: Crew Systems Directorate, Armstrong Laboratory, Air Force Materiel Command.

Greil, H., & Juergens, H. W. (2000). Variability of dimensions and proportions in adults or how to use classic anthropometry in man modeling. In K. Landau (Ed.), *Ergonomic software tools in product and workplace design: A review of recent developments in human modeling and other design aids* (pp.7–27). Stuttgart, Germany: Ergon.

Greiner, T. M., & Gordon, C. C. (1990). *An assessment of long-term changes in anthropometric dimensions: Secular trends of U.S. Army males* (NATICK/TR-91/006). Natick, MA: U.S. Army Natick Research, Development and Engineering Center.

Haas, E., & Edworthy, J. (Eds.). (2003). *The ergonomics of sound.* Santa Monica, CA: Human Factors and Ergonomics Society.

Haeger-Ross, C., & Roesblad, B. (2002). Norms for grip strength in children aged 4–16 years. *Acta Paediatrica, 91,* 617–625.

Haermae, M. (1996). Ageing, physical fitness and shiftwork tolerance. *Applied Ergonomics, 27,* 25–29.

Haeroe, J. M., & Vaerinen, S. (2003). Ergonomic approach for developing products aimed at older users: Method and product cases. In H. Strasser, K. Kluth, H. Rausch, & H. Bubb (Eds.), *Quality of work and products in enterprises of the future* (pp. 145–148). Stuttgart, Germany: Ergonomia Verlag.

Haex, B., & van Houte, R. (2001a). Sleeping systems: Current status. In W. Karwowski (Ed.), *International encyclopedia of ergonomics and human factors* (pp. 571–576). London: Taylor & Francis.

Haex, B., & van Houte, R. (2001b). Sleeping systems: Design requirements. In W. Karwowski (Ed.), *International encyclopedia of ergonomics and human factors* (pp. 566–570). London: Taylor & Francis.

Hamill, P. V. V., Drizd, T. A., Johnson, C. L., Reed, R. B., Roche, A. F., & Moore, W. M. (1979). Physical growth: National Center for Health statistics percentiles. *American Journal of Clinical Nutrition, 32,* 607–629.

Hammer, W., & Price, D. (2001). *Occupational safety management and engineering* (5th ed.). Upper Saddle River, NJ: Prentice-Hall.

Hancock, H. E., Fisk, A. D., & Rogers, W. A. (2001, Fall). Everyday products: Easy to use … or not? *Ergonomics in Design,* 12–18.

Hancock, P. A., & Desmond, P. A. (Eds.). (2001). *Stress, workload, and fatigue.* Mahwah, NJ: Erlbaum.

Hangartner, M. (1987). *Standardization in olfactometry with respect to odor pollution control: Assessment of odor annoyance in the community.* Presentations 87-75A.1 and 87-75B.3 at the 80th Annual Meeting of the APCA. New York: APCA.

Hayslip, B., & Panek, P. (1989). *Adult development and aging.* New York: Harper & Row.

Hedge, A., Barrero, M., & Maxwell, L. (2000). Ergonomic issues for classroom computing. In *Proceedings of the 14th Triennial Congress of the International Ergonomics Association and 44th Annual Meeting of the Human Factors and Ergonomics Society* (pp. 6.296–6.299). Santa Monica, CA: Human Factors and Ergonomics Society.

Helander, M. G. (2003). Forget about ergonomics in chair design? Focus on aesthetics and comfort! *Ergonomics, 46,* 1306–1319.

Helander, M. G., Landauer, T. K., & Prabhu, P. V. (Eds.). (1997). *Handbook of human-computer interaction* (2nd ed.). Amsterdam: Elsevier.

Helander, M. G., & Tham, M. P. (2003). Hedonomics — affective human factors design. *Ergonomics, 46,* 1269–1272.

Hertzberg, H. T. E. (1968). The conference on standardization of anthropometric techniques and terminology. *American Journal of Physical Anthropology, 28,* 1–16.

HFES 100 Committee. (2002). *Human factors engineering of computer workstations* (BSR/ HFES 100, Draft Standard for Trial Use). Santa Monica, CA: Human Factors and Ergonomics Society.

HFES 300 Committee. (2004). *Guidelines for using anthropometric data in product design.* Santa Monica, CA: Human Factors and Ergonomics Society.

Hirao, N., & Kajiyama, M. (1994) Seating for pregnant workers based on subjective symptoms and motion analysis. In R. Lueder & K. Noro (Eds.), *Hard facts about soft machines. The ergonomics of seating* (chap. 24, pp. 317–331). London: Taylor & Francis.

Ho, G., Scalfia, C. T., Caird, J. K., & Graw, T. (2001). Visual search for traffic signs: The effects of clutter, luminance, and aging. *Human Factors, 43,* 194–207.

Hochberg, F. H., Leffert, R. D., Heller, M. D., & Merriman, L. (1983). Hand difficulties among musicians. *Journal of the American Medical Association, 249*(14), 1869–1872.

Hsiao, H., Long, D., & Snyder, K. (2002). Anthropometric differences among occupational groups. *Ergonomics, 45,* 136–152.

Huang, C., & You, M. (1994). Anthropometry of Taiwanese women. *Applied Ergonomics, 25,* 186–187.

Huey, R. W., Buckley, D. S., & Lerner, N. D. (1994). Audible performance of smoke alarm sounds. In *Proceedings of the Human Factors and Ergonomics Society 38th Annual Meeting* (pp. 147–151). Santa Monica, CA: Human Factors and Ergonomics Society.

Hui, Y., Lee, L., Her, Y., & Tsuang, A (1999). Comparison of sitting posture adaptations of pregnant and non-pregnant females. *International Journal of Industrial Ergonomics, 23,* 391–396.

Ignazi, G., Martel, A., Mollard, R., & Coblentz, A. (1996). Anthropometric measurements. Evolution of a French schoolchildren and adolescent population aged 4 to 18. In *Proceedings of the 4th Pan Pacific Conference on Occupational Ergonomics* (pp. 111–114). Hsinchu, ROC: Ergonomics Society of Taiwan.

Imrhan, S. N. (1986). An analysis of finger pinch strength in children. In *Proceedings of the Human Factors Society 30th Annual Meeting* (pp. 667–671). Santa Monica, CA: Human Factors Society.

Imrhan, S. N. (2001). Physical strength in the older population. In W. Karwowski (Ed.), *International encyclopedia of ergonomics and human factors* (pp. 282–284). London: Taylor & Francis.

Imrhan, S. N., Nguyen, M. T., & Nguyen, N. N. (1993). Hand anthropometry of Americans of Vietnamese origin. *International Journal of Industrial Ergonomics, 12,* 281–287.

Intaranont, K. (1991). *Anthropometry and physical work capacity of agricultural and industrial populations in northeast Thailand. Report* (Laboratory for Ergonomics Research, Department of Industrial Engineering, Chulalongkorn University). Bangkok, Thailand: Chulalongkorn University Printing House. Abbreviated as Human characteristics of workers in northeast Thailand. In *Proceedings, 11th Congress of the International Ergonomics Association* (pp. 888–890). London: Taylor & Francis.

Jackson, A. S., Beard, E. F., Wier, L. T., & Stuteville, J. E. (1992). Multivariate model for defining changes in maximal physical working capacity of men, ages 25 to 70 years. In *Proceedings of the Human Factors Society 36th Annual Meeting* (pp. 171–174). Santa Monica, CA: Human Factors Society.

Jamieson, B. A., & Rogers, W. A. (2000). Age-related effects of blocked and random practice schedules on learning a new technology. *Journal of Gerontology: Psychological Sciences, 55B (6),* 343–363.

Jensen, R. C. (1999). Ergonomics in health-care organizations. In W. Karwowski & W. S. Marras (Eds.), *The occupational ergonomics handbook* (chap. 108, pp. 1949–1960). Boca Raton, FL: CRC Press.

Johnsson, C., Carlsson, R., & Lagerstroem, M. (2002). Evaluation of training in patient handling and moving skills among hospital and home care personnel. *Ergonomics, 45,* 850–865.

Juergens, H. W., Aune, I. A., & Pieper, U. (1990). *International data on anthropometry.* Occupational safety and health series #65. Geneva, Switzerland: International Labour Office.

Kagimoto, Y. (Ed.). (1990) *Anthropometry of JASDF personnel and its applications for human engineering.* Tokyo: Aeromedical Laboratory, Air Development and Test Wing JASDF.

Kalton, G., & Anderson, D. W. (1989). Sampling rare populations. In M. P. Lawton & A. R. Herzog (Eds.), *Special research methods for gerontology* (chap. 1, pp. 7–30). Amityville, NY: Baywood.

Kamentz, H. L. (1969). *The wheelchair book: Mobility for the disabled.* Springfield, IL: Charles C Thomas.

Kane, R. A., & Kane, R. L. (1981). *Assessing the elderly. A practical guide to measurement.* Lexington, MA: Lexington Books.

Karazman, R., Kloimueller, I., Gaertner, J, Geissler, H, Hoerwein, K., & Morawetz, I. K. (1998). Participatory development of age-adjusted, optional shift time schedules in industrial workers in a plant. In S. Kumar (Ed.), *Advances in occupational ergonomics and safety* (pp. 139–142). Amsterdam: IOS Press.

Karwowski, W. (Ed.) (2001). *International encyclopedia of ergonomics and human factors.* London: Taylor & Francis.

Karwowski, W., & Marras, W. S. (Eds.). (1999). *The occupational ergonomics handbook.* Boca Raton, FL: CRC Press.

Kayis, B., & Oezok, A. F. (1991a). The anthropometry of Turkish Army men. *Applied Ergonomics, 22,* 49–54.

Kayis, B., & Oezok, A. F. (1991b). Anthropometric survey among Turkish primary school-children. *Applied Ergonomics, 22,* 55–56.

Keir, P. J., & MacDonell, C. W. (2004). Muscle activity during patient transfers: A preliminary study on the influence of lift assists and experience. *Ergonomics, 47,* 296–306.

Kelly, P. L., & Kroemer, K. H. E. (1990). Anthropometry of the elderly: Status and recommendations. *Human Factors, 32,* 571–595.

Ketcham, C. J., & Stelmach, G. E. (2001). Age-related declines in motor control. In J. E. Birren & K. W. Schaie (Eds.), *Handbook of the psychology of aging* (5th ed, chap. 13, pp. 313–348). San Diego, CA: Academic.

Kira, A. (1976). *The bathroom.* New York: Viking.

Kline, D. W., & Scialfa, C. T. (1996). Visual and auditory aging. In J. E. Birren & K. W. Schaie (Eds.), *Handbook of the psychology of aging* (4th ed., chap. 10, pp. 181–203).). New York: Van Nostrand Reinhold.

Klinich, K. D., Schneider, L. W., Eby, B., Rupp, J., & Pearlman, M. D. (1999). *Seated anthropometry during pregnancy.* (UMTRI-99-16). Ann Arbor: University of Michigan, Transportation Research Institute.

Klinich, K. D., Schneider, L. W., Moore, J. L., & Pearlman, M. D. (2000). Investigation of crashes involving pregnant occupants. In *Proceedings of the Association for the Advancement of Automotive Medicine 44th Annual Conference* (pp. 37–55). Barrington, IL: Association for the Advancement of Automotive Medicine.

Knight, G., & Noyes, J. (1999). Children's behaviour and the design of school furniture. *Ergonomics 42,* 747–760.

Koga, T. (1994). The classification of physical disabilities in the design of products and spaces. In R. Lueder & K. Noro (Eds.), *Hard facts about soft machines: The ergonomics of seating* (chap. 23, pp. 299–316). London: Taylor & Francis.

Kondraske, G. V. (1988). Quantitative measurement and assessment of performance. In R. V. Smith & J. H. Leslie (Eds.), *Rehabilitation engineering.* Boca Raton, FL: CRC Press.

Konz, S., & Johnson, S. (2000). *Work design: Industrial ergonomics.* (5th ed.) Scottsdale, AZ: Holcomb Hathaway.

Koontz, A. M., Ambrosio, F., Souza, A. L., Buning, M. E., Arva, J., & Cooper, R. A. (2004). Strength and disability. In S. Kumar (Ed.), *Muscle strength* (chap. 21, pp. 485–518). Boca Raton, FL: CRC Press.

Koontz, T., & Dagwell, C. V. (1994). *Residential kitchen design.* New York: Van Nostrand Reinhold.

Koppa, R. J. (1990). State of the art in automotive adaptive equipment. *Human Factors, 32,* 439–455.

Kothiyal, K., & Tettey, S. (2000). Anthropometric data of elderly people in Australia. *Applied Ergonomics, 31,* 329–332.

Kramarow, E., Lentzner, H., Rooks, R., Weeks, J., & Saydah, S. (1999). *Health and aging chart book: Health, United States, 1999.* Hyattsville, MD: National Center for Health Statistics.

Kroemer, K. H. E. (1993, October; January 1994). Locating the computer screen: How high, how far? *Ergonomics in Design,* 7–8, 40.

Kroemer, K. H. E. (1997a). Anthropometry and biomechanics. In A. D. Fisk & W. A. Rogers (Eds.), *Handbook of human factors and the older adult* (chap. 5, pp. 87–124). San Diego, CA: Academic.

Kroemer, K. H. E. (1997b). *Ergonomic design of material handling systems*. Boca Raton, FL: CRC Press.

Kroemer, K. H. E. (1999a). Human strength evaluation. In W. Karwowski & W. S. Marras (Eds.), *The industrial ergonomics handbook* (chap. 11, pp. 205–227). Boca Raton, FL: CRC Press.

Kroemer, K. H. E. (1999b). Engineering anthropometry. In W. Karwowski & W. S. Marras (Eds.), *The occupational ergonomics handbook* (chap. 9, pp. 139–165). Boca Raton, FL: CRC Press.

Kroemer, K. H. E. (2001). Keyboards and keying. An annotated bibliography of the literature from 1878 to 1999. *International Journal Universal Access in the Information Society UAIS 1/2*, 99–160.

Kroemer, K. H. E. (2004). Design applications of strength data. In Kumar, S. (Ed.), *Muscle strength* (chap. 16, pp. 367–378). Boca Raton, FL: CRC Press.

Kroemer, K. H. E., & Grandjean, E. (1997a). *Fitting the task to the human* (5th ed.). London: Taylor & Francis.

Kroemer, K. H. E., & Kroemer, A. D. (2001a). *Office ergonomics*. London: Taylor & Francis.

Kroemer, K. H. E., & Kroemer, A. D. (2001b). Feeling good at work. In K. H. E. Kroemer & A. D. Kroemer (Eds.), *Office ergonomics* (chap. 6, pp. 129–191). London: Taylor & Francis.

Kroemer, K. H. E., Kroemer, H. B., & Kroemer-Elbert, K. E. (2001). *Ergonomics: How to design for ease and efficiency* (2nd. ed.). Upper Saddle River, NJ: Prentice-Hall.

Kroemer, K. H. E., Kroemer, H. J., & Kroemer-Elbert, K. E. (1997). *Engineering physiology: Bases of human factors/ergonomics* (3rd ed.). New York: Van Nostrand Reinhold–Wiley.

Kroemer, K. H. E., & Robinson, D. E. (1971). *Horizontal static forces exerted by men standing in common working positions on surfaces of various tractions* (AMRL-TR-70-114). Wright-Patterson Air Force Base, OH: Aerospace Medical Research Laboratory.

Kuczmarski, R. J., Ogden, C. L., Guo, S. S., Grummer-Strawn, L. M., Flegal, K. M., Mei, Z., Wei, R., Curtin, L. R., Roche, A. F., & Johnson, C. L. (2002). *2000 CDC growth charts for the United States: Methods and development.* (DHHS Publication No. PHS 2002-1696, Vital and Health Statistics, Series 11, No. 246). Hyattsville, MD: Department of Health and Human Services.

Kuenzi, J. K., & Kennedy, J.M. (1993). Anthropometry of surface mine maintenance workers. In W. S. Marras, W. Karwowski, & J. L. Smith (Eds.), *The ergonomics of manual work* (p. 203–206). London: Taylor & Francis.

Kumar, S. (Ed.). (2004). *Muscle strength*. Boca Raton, FL: CRC Press.

Kumashiro, M. (Ed.). (1995). *The paths to productive aging*. London: Taylor & Francis.

Kuorinka, I., & Forcier, L. (Eds.). (1995). *Work related musculoskeletal disorders: A reference book for prevention*. London: Taylor & Francis.

Kwahk, J., Smith-Jackson, T., & Williges, R. C. (2001). From user-centered design to senior-centered design: Designing Internet health information portals. In *Proceedings of the Human Factors and Ergonomics Society 45th Annual Meeting* (pp. 580–584). Santa Monica, CA: Human Factors and Ergonomics Society.

Laflamme, L., & Menckel, E. (1995). Aging and occupational accidents: Thirty years of conflicting findings. In M. Kumashiro (Ed.), *The paths to productive aging* (pp. 187–193). London: Taylor & Francis.

Laing, R. M., Holland, E. J., Wilson, C. A., & Niven, B. E. (1999). Development of sizing systems for protective clothing for the adult male. *Ergonomics, 42,* 1249–1257.

Lamey, J., Aghazadeh, F., & Nye, J. (1991). A study of the anthropometric measurements of Jamaican agricultural workers. In Y. Queinnec & F. Daniellou (Eds.), *Designing for everyone* (pp. 897–899). London: Taylor & Francis.

Landau, K. (Ed.). (2000). *Ergonomic software tools in product and workplace design. A review of recent developments in human modeling and other design aids.* Stuttgart, Germany: Ergon.

Lavender, S. A., Chen, S. H., Li, Y. C., & Andersson, G. B. J. (1998). Trunk muscle use during pulling tasks: Effects of a lifting belt and footing conditions. *Human Factors, 40,* 159–172.

Lawton, M. P. (1990). Aging and performance of home tasks. *Human Factors, 32,* 527–536.

Lawton, M. P., & Herzog, A. R. (Eds.). (1989). *Special research methods for gerontology.* Amityville, NY: Baywood.

Leboeuf-Yde, C. (2004). Back pain — Individual and genetic factors. *Electromyography and Kinesiology, 14 (Special Issue),* 129–133.

Le Bon, C., & Forrester, C. (1997). An ergonomic evaluation of a patient handling device: The elevate and transfer vehicle. *Applied Ergonomics, 28,* 365–374.

Lerner, N. D., & Huey, R. W. (1991). Residential fire needs of older adults. In *Proceedings of the Human Factors Society 35th Annual Meeting* (pp. 172–176). Santa Monica, CA: Human Factors and Ergonomics Society.

Levine, D. B., Zitter, M., & Ingram, L. (Eds.). (1990). *Disability statistics. An assessment* (Committee on National Statistics, National Research Council). Washington, DC: National Academy Press.

Liu, Y. (2003). Engineering aesthetics and aesthetic ergonomics: Theoretical foundations and a dual-process research methodology. *Ergonomics, 46,* 1273–1292.

Lockwood, A. H. (1989). Medical problems of musicians. *New England Journal of Medicine, 320*(4), 221–227.

Lohman, T. G., Roche, A. F., & Martorel, R. (Eds.). (1988). *Anthropometric standardization reference manual.* Champaign, IL: Human Kinetics.

Lowrey, G. H. (1986). *Growth and development of children* (8th ed.). Chicago: Year Book Medical Publishers.

Lueder, R., & Noro, K. (Eds.). (1994). *Hard facts about soft machines: The ergonomics of seating.* London: Taylor & Francis.

Lynch, R. M., & Freund, A. (2000). Short-term efficacy of back injury intervention project for patient care providers at one hospital. *American Industrial Hygiene Association Journal, 61,* 290–294.

Malassigne, P., & Amerson, T. L. (1992). In the era of ADA do we really have accessible bathrooms: A survey. In *Proceedings of the Human Factors Society 36th Annual Meeting* (pp. 578–581). Santa Monica, CA: Human Factors Society.

Malina, R. M., & Roche, A. F. (1983). *Manual of physical status and performance in childhood, Vol. 2: Physical performance.* New York: Plenum.

Marklin, R. W., & Simoneau, G. G. (2004). Design features of alternative computer keyboards: A review of experimental data. *Journal of Orthopedic Sports and Physical Therapy, 34,* 638–649.

Marquie, J. C., Cau-Bareille, D. P., & Volkoff, S. (Eds.). (1998). *Working with age.* London: Taylor & Francis.

Marquie, J. C., & Volkoff, S. (2001). Working with age: An ergonomic approach to aging on the job. In W. Karwowski (Ed.), *International encyclopedia of ergonomics and human factors* (pp. 613–616). London: Taylor & Francis.

Marras, W. S. (1999). Occupational biomechanics. In W. Karwowski & W. S. Marras (Eds.), *The industrial ergonomics handbook* (chap.10, pp. 167–204). Boca Raton, FL: CRC Press.

Marras, W. S. (2000). Occupational low back disorder causation and control. *Ergonomics, 43*, 880–902.

Marras, W. S., Davis, K. G., Ferguson, S. A., Lucas, B. R., & Purnendu, G. (2001). Spine loading characteristics of patients with low back pain compared with asymptomatic individuals. *Spine, 26*, 2566–2574.

Marras, W. S., Davis, K. G., Kirking, B. C., & Bertsche, P. K. (1999). A comprehensive analysis of low-back disorder risk and spinal loading during the transferring and repositioning of patients using different techniques. *Ergonomics, 42*, 904–926.

Marras, W. S., & Kim, J. Y. (1993). Anthropometry of industrial populations. *Ergonomics, 36*, 371–378.

Marras, W. S., & Waters, T. (Eds.). (2004). State of the art research perspectives on musculoskeletal disorder causation and control. *Electromyography and Kinesiology, 14 (Special Issue)*, 1–178.

McClelland, I. L., & Ward, J. S. (1982). The ergonomics of toilet seats. *Human Factors, 24*, 713–725.

McConville, J. T. (1978). Anthropometry in sizing and design. In NASA/Webb (Eds.), *Anthropometric sourcebook* (NASA Reference Publication 1024, Vol. 1, chap. 8, pp. 8.1–8.23). Houston: NASA Johnson Space Center.

McGill, S. M. (1999). Update on the use of back belts in industry: More data — same conclusion. In W. Karwowski & W. S. Marras (Eds.), *The occupational ergonomics handbook* (chap. 74, pp. 1353–1358). Boca Raton, FL: CRC Press.

McMillan, G. R. (2001). Brain and muscle signal-based control. In W. Karwowski (Ed.), *International encyclopedia of ergonomics and human factors* (pp. 379–381). London: Taylor & Francis.

McMillan, G. R., & Calhoun, G. L. (2001). Gesture-based control. In W. Karwowski (Ed.), *International encyclopedia of ergonomics and human factors* (pp. 237–239). London: Taylor & Francis.

McQuistion, L. (1993a). Ergonomics for one. *Ergonomics in Design, 1*, 9–10.

McQuistion, L. (1993b). Ergonomics-for-one: An introduction. In V. J. B. Rice (Ed.), *Ergonomics in health care and rehabilitation* (chap. 4, pp. 43–63). Boston: Butterworth-Heinemann.

Mead, S. E., Batsakes, P, Fisk, A. D., & Mykitsyshyn, A. (1999). Application of cognitive theory to training and design solutions for age-related computer use. *International Journal of Behavioral Development, 23*, 553–573.

Mead, S. E., Sit, R. A., Rogers, W. A., Jamieson, B. A., & Rousseau, G. K. (2000). Influences of general computer experience and age on library search performance. *Behaviour and Information Technology, 19*, 107–123.

Mebarki, B., & Davies, B. T. (1990). Anthropometry of Algerian women. *Ergonomics, 33*, 1537–1547.

Megaw, T. (2001). Published ergonomics literature. In W. Karwowski (Ed.), *International encyclopedia of ergonomics and human factors* (pp. 1920–1939). London: Taylor & Francis.

Milanese, S., & Grimmer, K. (2004). School furniture and the user population: An anthropometric perspective. *Ergonomics, 47*, 416–426.

Mital, A., & Karwowski, W. (Eds.). (1988). *Ergonomics in rehabilitation*. Philadelphia: Taylor & Francis.

Moelenbroek, J. F. M., Kroon-Ramaekers, Y. M. T., & Snijders, C. J. (2003). Revision of the design of a standard for the dimensions of school furniture. *Ergonomics, 46*, 681–694.

Moon, S. D., & Sauter, S. L. (Eds.). (1996). *Beyond biomechanics — Psychosocial aspects of musculoskeletal disorders in office work*. London: Taylor & Francis.

Moroney, W. F., & Cameron, J. A. (2001). An annotated review of selected military and government sources of human factors design related criteria. In W. Karwowski (Ed.), *International encyclopedia of ergonomics and human factors* (pp. 3–7). London: Taylor & Francis.

Morrell, R. W., & Echt, K. V. (1997). Designing written instructions for older adults: Learning to use computers. In A. D. Fisk & W. A. Rogers (Eds.), *Handbook of human factors and the older adult* (chap. 14, pp. 335–361). San Diego, CA: Academic.

Morris, N., Toms, M., Easthope, Y., & Biddulph, J. (1998). Mood and cognition in pregnant workers. *Applied Ergonomics, 29*, 377–381. See also the related 1999 Letter to the Editor by H. Gross, & H. Pattison in *Applied Ergonomics, 30*, 177.

Morrow, D., & Leirer, V. (1997). Aging, pilot performance, and expertise. In A. D. Fisk & W. A. Rogers (Eds.), *Handbook of human factors and the older adult* (chap. 9, pp. 199–230). San Diego, CA: Academic.

Mououdi, M.A. (1997). Static anthropometric characteristics of Tehran University students age 20–30. *Applied Ergonomics, 28,* 149–150.

Mououdi, M A., & Choobineh, A. R. (1997). Static anthropometric characteristics of students age range 6–11 in Mazandaran Province/Iran. *Applied Ergonomics, 28*, 145–147.

Moustafa, A. W., Davies, B. T., Darwich, M. S., & Ibraheem, M. A. (1987). Anthropometric study of Egyptian women. *Ergonomics, 30*, 1089–1098.

Mullick, A. (1997). Listening to people: Views about the bathroom. In *Proceedings of the Human Factors and Ergonomics Society 41st Annual Meeting* (pp. 500–504). Santa Monica, CA: Human Factors and Ergonomics Society.

Nag, P. K., Sebastian, N. C., & Mavlankar, M. G. (1980). Occupational work load of Indian agricultural workers. *Ergonomics, 23*, 91–102.

NASA. (1989). *Man-systems integration standards* (Revision A, NASA-STD 3000). Houston: NASA Johnson Space Center.

NASA/Webb. (Eds.). (1978). *Anthropometric sourcebook* (3 vols., NASA Reference Publication 1024). Houston: NASA Johnson Space Center.

National Research Council. (Ed.). (1979). *Odors from stationary and mobile sources.* Washington, DC: National Academy Press.

National Research Council. (Ed.). (1999). *Work-related musculoskeletal disorders: Report, workshop summary, and workshop papers.* Washington, DC: National Academy Press.

National Research Council and Institute of Medicine. (Eds.). (2001). *Musculoskeletal disorders and the workplace: Low back and upper extremities.* Washington, DC: National Academy Press.

Nayak, U. S. L. (1995, January). Elders-led design. *Ergonomics in Design, 8*–13.

Newell, A., & Gregor, P. (2000). Designing for extra-ordinary people and situations. *CSERIAC Gateway, 11(1), 12*–13.

Nicholls, J. A., & Grieve, D. W. (1992). Performance of physical tasks in pregnancy. *Ergonomics, 35*, 301–311.

Nickel, P., & Nachreiner, F. (2003). Sensitivity and diagnosticity of the 0.1-Hz component of heart rate variability as an indicator of mental workload. *Human Factors, 53*, 575–590.

Niebel, B. W., & Freivalds, A. (1999). *Methods, standards, and work design* (10th ed.). New York: McGraw-Hill.

Nordin, M., Andersson, G. B. J., & Pope, M. H. (1997). *Musculoskeletal disorders in the workplace: Principles and practices.* St. Louis, MO: Mosby.

Norris, B., & Wilson, J. R. (1995). *Childata: The handbook of child measurements and capabilities — Data for design safety* (DTI/Pub 1732/2k/6/2.96 AR). London: Department of Trade and Industry.

Noy, I. (2001). Visual perception, age, and driving. In W. Karwowski (Ed.), *International ency-clopedia of ergonomics and human factors* (pp. 348–349). London: Taylor & Francis.

Nussbaum, M. A., & Chaffin, D. B. (1996). Development and evaluation of a scaleable and deformable geometric model of the human torso. *Clinical Biomechanics, 11*, 25–34.

Nussbaum, M. A., & Torres, N. (2001). Effects of training in modifying working method during common patient-handling activities. *International Journal of Industrial Ergo-nomics, 27*, 33–41.

Ogawa, I., & Arai, K. (1995). Ergonomic considerations for Western-style toilets used in Japan. In *Proceedings of the International Ergonomics Association World Conference 1995* (pp. 745–748). Rio de Janeiro, Brazil: Associacao Brasiliera de Ergonomia.

Ogden, C. L., Fryar, C .D., Carroll, M. D., & Flegal, K. M. (2004). *Mean body weight, height, and body mass index, United States 1960–2002* (Advance Data from Vital and Health Statistics, No. 347, October 27, 2004). Hyattsville, MD: National Center for Health Statistics.

Olson, P. L. (2001). Driver perception and response. In W. Karwowski (Ed.), *International encyclopedia of ergonomics and human factors* (pp. 433–435). London: Taylor & Francis.

Ong, C. N., Koh, D., Phoon, W. O., & Low, A. (1988). Anthropometrics and display station preferences of VDU operators. *Ergonomics, 31*, 337–347.

Occupational Safety and Health Administration (OSHA). (Ed.). (2003). Guidelines for nursing homes. *Job Safety & Health Quarterly, 14* (3).

Ostlere, S. J., & Gold, R. H. (1991). Osteoporosis and bone density measurement methods. *Clinical Orthopaedics, 271*, 149–163.

Owings, C. L., Chaffin, D. B., Snyder, R. G., & Norcutt, R. (1975). *Strength characteristics of U.S. children for product safety design* (Report 011903-F). Ann Arbor: University of Michigan.

Pandolf, K. B. (1991). Aging and heat tolerance at rest or during work. *Experimental Aging Research, 17*, 189–204.

Panek, P. E. (1997). The older worker. In A. D. Fisk & W. A. Rogers (Eds.), *Handbook of human factors and the older adult* (chap. 15, pp. 363–394). San Diego, CA: Academic.

Paquet, E., Robinette, K. M., & Rioux, M. (2000) Management of three-dimensional and anthropometric databases: Alexandria and Cleopatra. *Journal of Electronic Imaging, 9*, 421–431.

Parsons, H. M. (1972). The bedroom. *Human Factors, 14, 421*–450.

Parsons, K. C. (2003). *Human thermal environments.* (3rd ed.) London: Taylor & Francis.

Paul, J. A., & Douwes, M. (1993). Two-dimensional photographic posture recording and description: A validity study. *Applied Ergonomics, 24*, 83–90.

Paul, J. A., Frings-Dresen, M. H. W. Salle, H. J. A., & Rozendal, R. H. (1995). Pregnant women and working surface height and working surface areas for standing manual work. *Applied Ergonomics, 26*, 129–133.

Peebles, L., & Norris, B. (1998). *Adultdata. The handbook of adult anthropometric and strength measurements — Data for design safety* (DTI/Pub 2917/3k/6/98/NP). London: Department of Trade and Industry.

Peebles, L., & Norris, B. (2000). *Strength data* (DTI/URN 00/1070). London: Department of Trade and Industry.

Peebles, L., & Norris, B. (2003). Filling "gaps" in strength data for design. *Applied Ergo-nomics, 34*, 73–88.

Peloquin, A. (1994). *Barrier-free residential design.* New York: McGraw-Hill.

Pennathur, A., Sivasubramaniam, S., & Contreras, L. R. (2003). Functional limitations in Mexican American elderly. *International Journal of Industrial Ergonomics, 31*, 41–50.

Perkins, T. C., & Blackwell, S. U. (1998). *Accommodation and occupational safety for pregnant military personnel* (AFRL-HE-WP-TR-1999-0019). Wright-Patterson Air Force Base, OH: Air Force Research Laboratory.

Peters, G. A., & Peters, B. J. (2001). The distracted driver. *Journal of the Royal Society for the Promotion of Health, 121(1),* 23–28.

Pheasant, S. (1996). *Bodyspace: Anthropometry, ergonomics and the design of work* (2nd ed.). London: Taylor & Francis.

Phillips, C. A. (2000). *Human factors engineering.* New York: Wiley.

Piccoli, B. (2001). Ergophthalmology: The visual system and work. In W. Karwowski (Ed.), *International encyclopedia of ergonomics and human factors* (pp. 212–218). London: Taylor & Francis.

Pinto, M. R., De Medici, S., Van Sant, C., Bianchi, A., Zlotniki, A., & Napoli, C. (2000). Ergonomics, gerontechnology, and design for the home-environment. *Applied Ergonomics, 31,* 317–322.

Pirkl, J. J. (Ed.). (1994). *Transgenerational design. Products for an aging population.* New York: Van Nostrand Reinhold.

Proos, L. A. (1993). Anthropometry in adolescence — Secular trends, adaption, ethnic and environmental differences. *Hormone Research, 39,* 18–24.

Putz-Anderson, V. (1988). *Cumulative trauma disorders: A manual for musculoskeletal diseases of the upper limbs.* London: Taylor & Francis.

Rappaport, M. (1970). Human factors applications in medicine. *Human Factors, 12,* 25–35.

Ray, G. G., Ghosh, S., & Atreya, V. (1995). An anthropometric survey of Indian schoolchildren aged 3–5 years. *Applied Ergonomics, 26,* 67–72.

Regnier, V. (1993). *Assisted living housing for the elderly: Design innovations from the United States and Europe.* New York: Wiley.

Regnier, V., Hamilton, J., & Yatabe, S. (1995). *Assisted living for the aged and frail.* New York: Columbia University Press.

Rice, V. J. B. (Ed.). (1998). *Ergonomics in health care and rehabilitation.* Boston: Butterworth-Heinemann.

Riihimaeki, H. (2001). Epidemiology of work-related back disorders. In F. Violante, T. Armstrong, & A. Kilbom (Eds.), *Occupational ergonomics: Work related musculoskeletal disorders of the upper limb and back* (chap. 2, pp. 11–19). London: Taylor & Francis.

Roberto, K. A. (2001a). Chronic pain and intimacy in the relationships of older adults. *Generations, 25,* 65–74.

Roberto, K. A., Allen, K., & Blieszner, R. (2001b). Older adults' preferences for future care: Familial support versus formal plans. *Applied Developmental Sciences, 5,* 112–120.

Roberto, K. A., & Gold, D. T. (2001c). *Chronic pain in later life: A selectively annotated bibliography.* Westport, CT: Greenwood.

Robinette, K. M. (1998) Multivariate methods in engineering anthropometry. In *Proceedings of the Human Factors and Ergonomics Society 42nd Annual Meeting* (pp. 719–721). Santa Monica, CA: Human Factors and Ergonomics Society.

Robinette, K. M. (2000). CAESAR measures up. *Ergonomics in Design, 8,* 17–23.

Robinette, K. M., Blackwell, S, Daanen, H., Boehmer, M, Fleming, S., Brill, T., Hoeferlin, D., & Burnsides, D. (2002a). *Civilian American and European surface anthropometry resource (CAESAR) final report, Vol. I: Summary* (AFRL-HE-WP-TR-2002-0169). Wright-Patterson Air Force Base, OH: Air Force Research Laboratory.

Robinette, K. M., Blackwell, S, Daanen, H., Boehmer, M, Fleming, S., Brill, T., Hoeferlin, D., & Burnsides, D. (2002b). *Civilian American and European surface anthropometry resource (CAESAR) final report, Vol. II: Descriptions* (AFRL-HE-WP-TR-2002-0173). Wright-Patterson Air Force Base, OH: Air Force Research Laboratory.

Robinette, K. M., & Daanen, H. (2003). Lessons learned from CAESAR: A 3-D anthropometric survey. In *Proceedings of the 15th Triennial Congress of the International Ergonomics Association (Paper No. 00730).* London: Taylor & Francis.

Roche, A. F., & Malina, R. M. (1983). *Manual of physical status and performance in childhood, Vol. 1: Physical status.* New York: Plenum.

Roebuck, J. A. (1995). *Anthropometric methods: Designing to fit the human body.* Santa Monica, CA: Human Factors and Ergonomics Society.

Rogers, W. A. (1997a). Individual differences, aging, and human factors: An overview. In A. D. Fisk & W. A. Rogers (Eds.), *Handbook of human factors and the older adult* (chap. 7, pp. 155–170). San Diego, CA: Academic.

Rogers, W. A. (Ed.). (1997b). *Designing for an aging population: Ten years of human factors/ergonomics research.* Santa Monica, CA: Human Factors and Ergonomics Society.

Rogers, W. A., & Fisk, A. D. (Eds.). (2000). Human factors applied cognition, and aging. In F. I. M. Craik & T. A. Salthouse (Eds.), *The handbook of aging and cognition* (2nd ed., chap. 10, pp. 559–591).). Mahwah, NJ: Erlbaum.

Rogers, W. A., & Fisk, A. D. (Eds.). (2001). *Human factors interventions for the health care of older adults.* Mahwah, NJ: Erlbaum.

Rogers, W. A., Meyer, B., Walker, N., & Fisk, A. D. (1998). Functional limitations to daily living tasks in the aged: A focus group analysis. *Human Factors, 40,* 111–125.

Rogers, W. A., Mykityshin, A. L., Campbell, R. H., & Fisk, A. D. (2001). Analysis of a "simple" medical device. *Ergonomics in Design, 9,* 6–14.

Roush, L., & Koppa, R. (1992). A survey of activation importance of individual secondary controls in modified vehicles. *Assistive Technology, 4,* 66–69.

Rutter, B. G., Haager, J. A., Daigle, G. C., Smith, S., McFarland, N., & Kelsey, N. (1984). Dimensional changes throughout pregnancy: A preliminary report. *Carle Select Papers, 36,* 44–52.

Saito, S., Sotoyama, M., Jonai, H., Akutsu, M., Yatani, M., & Marumoto, T. (2000). Research activities on the ergonomics of computers in schools in Japan. In *Proceedings of the 14th Triennial Congress of the International Ergonomics Association and 44th Annual Meeting of the Human Factors and Ergonomics Society* (p. 7-658). Santa Monica, CA: Human Factors and Ergonomics Society.

Salthouse, T. A. (2000). Methodological assumptions in cognitive aging research. In F. I. M. Craik & T. A. Salthouse (Eds.), *The handbook of aging and cognition* (2nd ed., pp. 467–498). Mahwah, NJ: Erlbaum.

Salvendy, G. (Ed.). (1997). *Handbook of human factors and ergonomics* (2nd ed.). New York: Wiley.

Scherer, M., & Lane, J. (1997). Assessing consumer profiles of "ideal" assistive technologies in ten categories: An integration of quantitative and qualitative methods. *Disability and Rehabilitation, 19,* 528–535.

Schibye, B., Hansen, A. F., Hye-Knudsen, C. T., Essendrop, M., Boecher, M., & Skotte, J. (2003). Biomechanical analysis of the effect of changing patient-handling technique. *Applied Ergonomics, 34,* 115–123.

Schieber, F. (1994). Age and glare recovery time for low-contrast stimuli. In *Proceedings of the Human Factors and Ergonomics Society 38th Annual Meeting* (pp. 496–499). Santa Monica, CA: Human Factors and Ergonomics.

Schlosser, E. (2002). *Fast food nation* (rev. ed., pp. 121–129). New York: Perennial/HarperCollins.

Schmidt, R. F., & Thews, G. (Eds.). (1989). *Human physiology* (2nd ed.). New York: Springer.

Schneider, L. W., Klinich, K. D., Rupp, J., Moore, J., Eby, B., Ashton-Miller, J. A., Moss, S., Zhou, J., & Pearlman, M. D. (2000). Improving automotive safety during pregnancy. *UMTRI Research Review 31*(1), 1–11.

Schneider, L. W., Lehman, R. J., Pflug, M. A., & Owings, C. L. (1986). *Size and shape of the head and neck from birth to four years* (Final Report CPSC-C-83-1250). Ann Arbor: University of Michigan, Transportation Research Institute.

Scholey, M., & Hair, M. (1989). Back pain in physiotherapists involved in back care education. *Ergonomics, 32,* 179–190.

Schulz, R., Czaja, S. J., & Belle, S. (2001). Overview and intervention approaches to family care giving: Decomposing complex psychosocial interventions. In W. A. Rogers & A. D. Fisk (Eds.), *Human factors interventions for the health care of older adults* (chap. 8, pp. 147–164). Mahwah, NJ: Erlbaum.

Seitz, T., Garcia-Gil, H., Stuedeli, T., & Menozzi, M. (2003). Digital image processing for the determination of body movements in young children. In H. Strasser, K. Kluth, H. Rausch, & H. Bubb (Eds.), *Quality of work and products in enterprises of the future* (pp. 83–86). Stuttgart, Germany: Ergonomia Verlag.

Selkoe, D. J. (1992). Aging brain, aging mind. *Scientific American, 267,* 135–142.

Serow, W. J., & Sly, D. F. (1988). The demography of current and future aging cohorts. In Committee on an Aging Society (Ed.), *America's aging — The social and built environment in an older society* (pp. 42–102). Washington, DC: National Academy Press.

Shephard, R. J. (1995). A personal perspective on aging and productivity, with particular reference to physically demanding work. *Ergonomics, 38,* 617–636.

Sherrick, C. E., & Cholewiak, R. W. (1986). Cutaneous sensitivity. In K. R. Boff, L. Kaufmann, & J. P. Thomas (Eds.), *Handbook of perception and human performance* (chap. 12, pp. 12.1–12.58). New York: Wiley.

Sholes, C. L (1878). Improvement in type-writing machines. *Letters Patent No. 207,559.* United States Patent Office.

Siervogel, R. M., Roche, A. F., Guo, S., Mukherjee, D., & Chumlea, W. C. (1991). Patterns of change in weight/stature from 2 to 18 years: Findings from long-term serial data for children in the Fels longitudinal growth study. *International Journal of Obesity, 15,* 479–485.

Silverstein, B., & Clark, R. (2004). Interventions to reduce work-related musculoskeletal disorders. *Electromyography and Kinesiology, 14 (Special Issue),* 135–152.

Singh, J., Pen, C. M., Lim, M. K., & Ong, C. N. (1995). An anthropometric study of Singapore candidate aviators. *Ergonomics, 38,* 651–658.

Small, A. M. (1997). Design for older people. In G. Salvendy (Ed.), *Handbook of human factors and ergonomics* (2nd ed., pp. 495–504). New York: Wiley.

Smith, D. B. D. (1990). Human factors and aging. An overview of research needs and application opportunities. *Human Factors, 32,* 509–526.

Smith, R. V., & Leslie, J. H. (Eds.). (1990). *Rehabilitation engineering.* Boca Raton, FL. CRC Press.

Smith, S., Norris, B., & Peebles, L. (2000). *Older adultdata. The handbook of measurements and capabilities of the older adult — data for design safety* (DTI Pub 4445/3k/01/00/NP. ISBN 0/9522571/57). London: Department of Trade and Industry.

Smith-Jackson, T. L., & Williges, R. C. (2001). User-centered design of telesupport systems. *Assistive Technology, 13,* 144–169.

Smolowe, J. (1996). Older, longer. *Time (Special Fall Issue),* 76–80.

Snook, S. H. (2001). Back risk factors: An overview. In F. Violante, T. Armstrong, & A. Kilbom (Eds.), *Occupational ergonomics: Work related musculoskeletal disorders of the upper limb and back* (chap. 11, pp. 129–148). London: Taylor & Francis.

Snook, S. H. (2004). Work-related low back pain: Secondary intervention. *Electromyography and Kinesiology, 14 (Special Issue),* 153–160.

Snook, S. H., Webster, B. S., & McGorry, R. W. (2002). The reduction of chronic, non-specific low back pain through the control of early morning lumbar flexion: Three-year follow-up. *Journal of Occupational Rehabilitation, 12*, 13–19.

Snyder, R. G., Schneider, L. W., Owings, C. L., Reynolds, H. M., Golomb, D. H., & Schork, M. A. (1977). *Anthropometry of infants, children, and youths to age 18 for product safety design* (Final Report UM-HSRI-88-17). Ann Arbor: University of Michigan, Highway Safety Research Institute.

Snyder, R. G., Spencer, M. L., Owings, C. L., & Schneider, L. W. (1975). *Physical characteristics of children as related to death and injury for consumer product safety design* (Final Report, UM-HSRI-BI-75-5). Ann Arbor: University of Michigan, Highway Safety Research Institute.

Soldo, B. J., & Longino, C. F. (1988). Social and physical environments for the vulnerable aged. In Committee on an Aging Society (Ed.), *America's aging — The social and built environment in an older society* (pp. 103–133). Washington, DC: National Academy Press.

Sommerich, C. M., Joines, S. M. B., & Psihogios, J. P. (2001). Effects of computer monitor viewing angle and related factors on strain, performance, and preference outcomes. *Human Factors, 43*, 39–55.

Spence, A. P. (1994). *Biology of human aging* (2nd ed.). Englewood Cliffs, NJ: Prentice Hall.

Staffel, F. (1884). On the hygiene of sitting (in German). *Zbl. Allgemeine Gesundheitspflege, 3*, 403–421.

Staffel, F. (1889). *The types of human postures and their relations to deformations of the spine* (in German). Wiesbaden, Germany: Bergmann.

Stanton, N. A., & Young, M. S. (1999). *A guide to methodology in ergonomics*. London: Taylor & Francis.

Steenbekkers, L. P. A. (1993). *Child development, design implications and accident prevention*. Delft: Delft University Press.

Steenbekkers, L. P. A., & Beijsterveldt, C. E. M. (Eds.). (1998). *Design-relevant characteristics of ageing users*. Delft: Delft University Press.

Steenbekkers, L. P. A., & Molenbroek, J. F. M. (1990). Anthropometric data of children for non-specialist users. *Ergonomics, 33*, 421–429.

Steinfeld, E. (Ed.). (2001). *Anthropometrics of disability: An international workshop* (RERC on Universal Design at Buffalo). Buffalo, NY: University at Buffalo, School of Architecture and Planning.

Steinfeld, E., Schroeder, S., Duncan, J., Wirth, P., & Faste, R. (1979). *Access to the built environment: A review of literature* (PB80-136526, U.S. Department of Commerce, Dept. of Housing and Urban Development). Washington, DC: Superintendent of Documents.

Stoudt, H. W. (1981). The anthropometry of the elderly. *Human Factors, 23*, 29–37.

Straker, L., Harris, C., & Zandvliet, D. (2000). Scarring a generation of schoolchildren through poor introduction to technology in schools. In *Proceedings of the 14th Triennial Congress of the International Ergonomics Association and 44th Annual Meeting of the Human Factors and Ergonomics Society* (pp. 6.300–6.303). Santa Monica, CA: Human Factors and Ergonomics Society.

Strauss, R. S., & Pollack, H. A. (2001). Epidemic increase in childhood overweight, 1986–1998. *Journal of the American Medical Association, 22*, 2845–2848.

Strokina, A. N., & Pakhomova, B. A. (1999). *Anthropo-ergonomic atlas* (in Russian). Moscow: Moscow State University Publishing House (ISBN 5-211-04102-X).

Swanson, A. B., Goran-Hagert, C., & Swanson, G. D. G. (1987). Evaluation of impairment in the upper extremity. *Journal of Hand Surgery, 12A*, No. 5, Part 2, 896–926.

Swink, J. R. (1966). Intersensory comparisons of reaction time using an electro-pulse tactile stimulus. *Human Factors, 8*, 143–145.

Tayyari, F., & Smith, J. L. (1997). *Occupational ergonomics — Principles and applications.* New York: Chapman and Hall.

Tenner, E. (2003a). Letter perfect? Text keyboards. In *Our own devices: The past and future of body technology* (chap. 8, pp 187–212). New York: Knopf.

Tenner, E. (2003b). Sitting up straight and laid back: Reclining chairs. In *Our own devices: The past and future of body technology* (chap. 5, pp. 104–133; chap. 6, pp. 134–160). New York: Knopf.

Theilmeier, A., Jordan, C., Jaeger, M., & Luttmann, A. (2003). Measurement of exerted force during patient transfer for determining lumbar load. In H. Strasser, K. Kluth, H. Rausch, & H. Bubb (Eds.), *Quality of work and products in enterprises of the future* (pp. 1006–1009). Stuttgart, Germany: Ergonomia Verlag.

Thoumier, P., Drape, J. L., Aymard, C., & Bedoiseau, M. (1998). Effects of a lumbar support on spine posture and motion assessed by electrogoniometer and continuous recording. *Clinical Biomechanics, 13*, 18–26.

Treaster, D. E., & Marras, W. S. (2000). An assessment of alternate keyboards using finger motion, wrist motion and tendon travel. *Clinical Biomechanics, 15*, 499–503.

Triolo, R. J., Reilley, W. B., Freedman, W., & Bertz, R. R. (1993). Development and standardization of a clinical evaluation of standing function. The functional standing test. *IEEE Transactions on Rehabilitation Engineering, 1*(1), 18–25.

Troussier, B., Tesniere, C., Fauconnier, J., Grison, J., Juvin, R., & Phelip, X. (1999). Comparative study of two different kinds of school furniture among children. *Ergonomics, 42*, 516–526.

Troy, B. S., Cooper, R. A., Robertson, R. N., & Grey, T. L. (1995). Analysis of work postures of manual wheelchair users. In *Proceedings of the ErgoCon '95, Silicon Valley Ergonomics Conference and Exposition* (pp. 166–171). San Jose, CA: San Jose State University.

Tsunawake, N., Tahara, Y., Yukawa, K., Katsuura, T., Harada, A. Iwanaga, K., & Kikuchi, Y. (1995). Changes in body shape of young individuals from the aspect of adult physique model by factor analysis. *Applied Human Science, 14*, 227–234.

Ulin, S. S., Chaffin, D. B., Patellos, C. L., Blitz, S. G., Emerick, C. A., Lundy, F., & Misher, L. (1997). A biomechanical analysis of methods used for transferring totally dependent patients. *SciNursing, 14*, 19–27.

University of Nottingham. (2002). *Strength data for design safety, Phase 2* (DTI URN 01/1433). London: Department of Trade and Industry.

U.S. Army. (1981). *MIL-HDBK 759. Human factors engineering design for army material (metric).* Redstone Arsenal, AL: U.S. Army Missile Command.

Vanderheiden, G. C. (1997). Design for people with functional limitations resulting from disability, aging, or circumstance. In G. Salvendy (Ed.), *Handbook of human factors and ergonomics* (2nd ed., chap. 60, 2010–2052). New York: Wiley.

Van der Woude, L. G. V., Meijs, P. J. M., & de Boer, Y. A. (Eds.). (1991). *Wheelchair workshop workbook.* Amsterdam: Vrije Universiteit.

Verbrugge, L. M. (1991). Physical and social disability in adults. In H. Hibbard, P. A. Nutting, & M. L. Grady (Eds.), *Primary care research: Theory and methods* (pp. 31–57, AHCPR Publication No. 91-0011). Rockville, MD: U.S. Department of Health and Human Services.

Vercruyssen, M. (1997). Movement control and speed of behavior. In A. D. Fisk & W. A. Rogers (Eds.), *Handbook of human factors and the older adult* (chap. 4, pp. 55–86). San Diego, CA: Academic.

Vicente, K. J. (2002). Ecological interface design: Progress and challenges. *Human Factors, 43,* 62–78.

Victor, V. M., Nath, S., & Verma, A. (2002). Anthropometric survey of Indian farm workers to approach ergonomics in agricultural machinery design. *Applied Ergonomics, 33,* 579–581.

Violante, F., Isolani, I., & Raffi, G. B. (2000). Case definition for upper limb disorders. In F. Violante, T. Armstrong, & A. Kilbom (Eds.), *Occupational ergonomics. Work related musculoskeletal disorders of the upper limb and back* (chap. 10, pp. 120–128). London: Taylor & Francis.

Wang, M. J. J., Wang, E. M. Y., & Lin, Y. C. (2002a). *Anthropometric data book of the Chinese people in Taiwan.* Hsinchu, ROC: Ergonomics Society of Taiwan.

Wang, M. J. J., Wang, E. M. Y., & Lin, Y. C. (2002b). The anthropometric database for children and young adults in Taiwan. *Applied Ergonomics, 6,* 583–585.

Wang, Y., Wang, J. Q., Hesketh, T., Dang, Q. J., Mulligan, J., & Kinra, S. (2000, November 4). Standard definition of child overweight and obesity (letter to the editor). *British Medical Journal, 321,* 1158–1159.

Wargo, M. J. (1967). Human operator response speed, frequency and flexibility: A review and analysis. *Human Factors, 9,* 221–238.

Webb, P. (1985). *Human calorimeters.* New York: Praeger.

Weimer, J. (Ed.). (1995). *Research techniques in human engineering.* Englewood Cliffs, NJ: Prentice Hall.

Wickens, C. D., Gordon, S. E., & Liu, Y. (1998). *An introduction to human factors engineering.* New York: Longman.

Wiklund, M. E., & Smith, L. (1992). New Human Factors Society technical group: Medical systems and functionally impaired persons. *CSERIAC Gateway, 3*(3), 6.

Wilkoff, W. L., Wilkoff, P. C., & Abed, L. W. (1998). Complying with the Americans with Disabilities Act: A design retrofit. In V. J. B. Rice (Ed.), *Ergonomics in health care and rehabilitation* (chap. 20, pp. 335–351). Boston: Butterworth-Heinemann.

Williges, R. C., & Williges, B. H. (1995). Travel alternatives for the mobility impaired: The surrogate electronic traveler (SET). In A. D. N. Edwards (Ed.), *Extra-ordinary human–computer interaction: Interfaces for users with disabilities* (chap. 12, pp. 245–262). New York: Cambridge University Press.

Wilson, G. F., & Russell, A. (2003). Real-time assessment of mental workload using psycho-physiological measures and artificial neural networks. *Human Factors, 45,* 635–643.

Winter, D. A. (1990). *Biomechanics and motor control of human movement* (2nd ed.). New York: Wiley.

Winter, S. (1997). *Accessible housing by design: Universal design principles in practice.* New York: McGraw-Hill.

Wood, J. M. (2002). Age and visual impairment decrease driving performance as measured on a closed-road circuit. *Human Factors, 43,* 482–494.

Woods, D. D. (2000) Patient safety and human factors opportunities. *Human Factors and Ergonomics Society Bulletin, 43,* 1, 4, 5.

Wright, W. C. (1993). *Diseases of workers.* Translation of Bernadino Ramazzini's *1713 De Morbis Articum.* Thunder Bay, Ontario, Canada: OH&S Press.

Wu, G., Siegler, S., Allard, P., Kirtley, C., Leardini, A., Rosenbaum, D., Whittle, M., Lima, D. D., Cristofolini, L., Witte, H., Schmid, O., & Stokes, I. (2002). ISB recommendation on definitions of joint coordinate system of various joints for the reporting of human joint motion. Part I: Ankle, hip, and spine. *Journal of Biomechanics, 35,* 543–555.

Wylde, M., Baron-Robbins, A., & Clark, S. (1994). *Building for a lifetime.* Newtown, CT: Taunton.

Yadav, R., Tewari, V.K., & Prasad, N. (1997). Anthropometric data of Indian farm workers — A module analysis. *Applied Ergonomics, 28,* 69–71.

Young, A. J. (1991). Effects of aging on human cold tolerance. *Experimental Aging Research, 17,* 205–213.

Yu, T., Roht, L. H., Wise, R. A., Kilian, D. J., & Weir, F. W. (1984). Low-back pain in industry: An old problem revisited. *Journal of Occupational Medicine, 26,* 517–524.

Zacharkow, D. (1984). *Wheelchair posture and pressure sores.* Springfield. IL: Charles C Thomas.

Zacharkow, D. (1988). *Posture: Sitting, standing, chair design and exercise.* Springfield, IL. Charles C Thomas.

Zhuang, Z., Stobbe, T. J., Collins, J. W., Hsiao, H., & Hobbs, G. R. (2000). Psychophysical assessment of assistive devices for transferring patients/residents. *Applied Ergonomics, 31,* 35–44.

Appendix

The following tables present anthropometric information on "normal" adults in various regions of the globe. Chapters 3 and 4 contain discussions of anthropometric procedures and statistics.

Table A.1 summarizes the data from all published surveys that came to my attention, with no specific selection criteria applied.

Tables A.2 through A.10 provide information from recent surveys that I selected because they appear reasonably representative and well conducted. I arranged the data in a common format to make comparisons easy. Disappointingly, there are only nine such data sets. Apparently, current and comprehensive anthropometric surveys are not abundant.

When comparing the data in the following tables (including those estimated by Juergens and coauthors in 1990, shown as Table 1.1 in the first chapter), consider that the surveys vary in dates and sample sizes. In most cases, little information exists about the local selection of the measured people.

In spite of these concerns, these tables can serve as benchmarks for comparisons with the results of measurements on samples of "extra-ordinary" people.

TABLE A.1
International Anthropometric Measured Data — Averages
(with Standard Deviations)

Country	Sample Size	Stature mm	Sitting Height mm	Knee Height, Sitting mm	Weight kg
Algeria:					
Females (Mebarki & Davies, 1990)	666	1576 (56)	795 (50)	487 (36)	61 (1)
Australia:					
Females, 77 (8) years old	138	1521 (70)	775 (40)	—	61 (13)
Males, 76 (7) years old (Kothiyal & Tettey, 2000)	33	1658 (79)	843 (56)	—	72 (11)
Brazil:					
Males (Ferreira, 1988; cited by M. Al-Haboubi, 1991)	3076	1699 (67)	—	—	—
China:					
Females (Hong Kong)	69	1607 (54)	838 (45)	510 (31)	—
Females (Taiwan) (Huang & You, 1994)	300	1582 (49)	—	—	51 (7)
Females (Taiwan)	about 600**	1572 (53)	846 (32)	471 (24)	52 (7)
Males (Hong Kong) (Chan, So, & Ng, 2000)	286	1737 (49)	884 (42)	552 (29)	—
Males (Canton) (Evans, 1990)	41	1720 (63)	—	—	60 (6)
Males (Taiwan) (Wang, Wang, & Lin, 2000)	about 600**	1705 (59)	910 (30)	521 (29)	67 (9)
Egypt:					
Females (Moustafa, Davies, Darwich, & Ibraheem, 1987)	4960	1606 (72)	838 (43)	499 (25)	63 (4)
France:					
Females	328	1620	867	487	58
Males (Coblentz, personal communication, 1997)	687	1747	918	533	70
Germany (East):					
Females	123	1608 (59)	854 (31)	497 (24)	—
Males (Fluegel, Greil, & Sommer, 1986)	30	1715 (66)	903 (34)	531 (27)	—
India:					
Females	251	1523 (66)	775 (39)	483 (28)	50 (10)
Males (Chakarbarti, 1997)	710	1650 (70)	937 (45)	520 (30)	57 (11)
Central India male farm workers (Gite & Yadav, 1989)	39	1620 (50)	739 (26)	509 (30)	49 (6)
East–Central India male farm workers (Victor, Nath, & Verma, 2002)	300	1638 (56)	775 (40)	—	57 (7)
South India male workers (Fernandez & Uppugonduri, 1992)	128	1607 (60)	791 (40)	542 (38)	57 (5)

TABLE A.1 (continued)
International Anthropometric Measured Data — Averages
(with Standard Deviations)

Country	Sample Size	Stature mm	Sitting Height mm	Knee Height, Sitting mm	Weight kg
India:					
East India male farm workers	134	1621 (58)	809 (22)	515 (29)	54 (67)
(Yadav, Tewari, & Prasad, 1997)					
Indonesia:					
Females	468	1516 (54)	719 (34)	—	—
Males (Sama'mur, 1985; cited by	949	1613 (56)	872 (37)	—	—
Intaranont, 1991)					
Iran:					
Female students	74	1597 (58)	861 (36)	488 (23)	56 (10)
Male students	105	1725 (58)	912 (26)	531 (24)	66 (10)
(Mououdi, 1997)					
Ireland:					
Males	164	1731 (58)	911 (30)	508 (28)	74 (9)
(Gallwey & Fitzgibbon, 1991)					
Italy:					
Females	753*	1610 (64)	850 (34)	495 (30)	58 (8)
Femailes	386**	1611 (62)	—	—	58 (9)
Males	913*	1733 (71)	896 (36)	541 (30)	75 (10)
Males	410**	1736 (67)	—	—	73 (11
*(Coniglio, Inbini, Mascal et al., 1991)					
**(Robinette, Blackwell, Daanen et al., 2002)					
Jamaica:					
Females	123	1648	832	—	61
Males	30	1749	856	—	68
(Lamey, Aghazadeh, & Nye, 1991)					
Japan:					
Females	240	1584 (50)	855 (28)	475 (20)	54 (6)
Males (Kagimoto, 1990)	248	1688 (55)	910 (30)	509 (22)	66 (8)
Korea (South):					
Female workers (Fernandez,	101	1580 (57)	833 (32)	460 (22)	54 (7)
Malzahn, Eyada, & Kim, 1989)					
Malaysia:					
Females	32	1559 (66)	831 (39)	—	—
(Ong, Koh, Phoon, & Low, 1988)					
Netherlands:					
Females, 20–30 yrs	68*	1686 (66)	—	—	67 (10)
Females, 18–65 yrs	691**	1679 (75)	—	—	73 (16)
Males, 20–30 yrs	55*	1848 (80)	—	—	81 (14)
Males, 18–65 yrs	564**	1813 (90)	—	—	84 (16)
*(Steenbekkers & Beijsterveldt, 1998)					
**(Robinette, Blackwell, Daanen et al., 2002a)					

TABLE A.1 (continued)
International Anthropometric Measured Data — Averages
(with Standard Deviations)

Country	Sample Size	Stature mm	Sitting Height mm	Knee Height, Sitting mm	Weight kg
Russia:					
Female herders (ethnic Asians)	246	1588 (55)	—	—	—
Female students (ethnic Russians)	207	1637 (57)	859 (32)	527 (24)	61 (8)
Female students (ethnic Uzbeks)	164	1578 (49)	839 (28)	487 (25)	56 (7)
Female factory workers (ethnic Russians)	205	1606 (53)	849 (30)	494 (26)	61 (8)
Female factory workers (ethnic Uzbeks)	301	1580 (54)	845 (31)	484 (26)	58 (9)
Male students (ethnic Russians)	166	1757 (56)	912 (32)	562 (25)	71 (9)
Male students (ethnic Uzbeks)	150	1700 (52)	905 (29)	531 (23)	65 (7)
Male factory workers (ethnic Russians)	192	1736 (61)	909 (32)	550 (25)	72 (10)
Male factory workers (ethnic mix)	150	1700 (59)	896 (32)	541 (24)	68 (8)
Male farm mechanics (ethnic Asians)	520	1704 (58)	902 (31)	530 (25)	64 (8)
Male coal miners (ethnic Russians)	150	1801 (61)	978 (33)	572 (25)	—
Male construction workers (ethnic Russians) (Strokina & Pakhomova, 1999)	150	1707 (69)	—	—	—
Saudi Arabia:					
Males (Dairi, 1986; cited by Al-Haboubi, 1991)	1440	1675 (61)	—	—	—
Singapore:					
Females (Ong, Koh, Poon, & Low, 1988)	46	1598 (58)	855 (31)	—	—
Males (pilot trainees) (Singh, Peng, Lim, & Ong, 1995)	832	1685 (53)	894 (32)	—	—
Sri Lanka:					
Females	287	1523 (59)	774 (22)	—	—
Males (Abeysekera, 1985; cited by Intaranont, 1991)	435	1639 (63)	833 (27)	—	—
Sudan:					
Males:					
Villagers	37*	1687 (63)	—	—	57 (8)
City dwellers	16*	1704 (72)	—	—	62 (13)
	48**	1668	—	—	51
Soldiers					
*(ElKarim, Sukkar, Collins, & Doré, 1981)	21*	1735 (71)	—	—	71 (8)
**(Ballal et al., 1982; cited by Intaranont, 1991	104**	1728	—	—	60

TABLE A.1 (continued)
International Anthropometric Measured Data — Averages
(with Standard Deviations)

Country	Sample Size	Stature mm	Sitting Height mm	Knee Height, Sitting mm	Weight kg
Thailand:					
Females	250*	1512 (48)	—	—	—
	711*	1540 (50)	817 (27)	—	—
Males *(Intaranont, 1991)	250*	1607 (20)	—	—	—
**(NICE; cited by Intaranont, 1991)	1478**	1654 (59)	872 (32)	—	—
Turkey:					
Females:					
Villagers	47	1567 (52)	792 (38)	486 (27)	69 (14)
City dwellers (Goenen, Kalinkara, & Oezgen, 1991)	53	1563 (55)	786 (05)	471 (05)	66 (13)
Male soldiers (Kayis & Oezok, 1991)	5108	1702 (60)	888 (34)	513 (28)	63 (7)
United States:					
Females	about 3800	1625	—	—	75
Males	about 3800	1762	—	—	87
(Ogden, Fryar, Carroll et al., 2004)					
North American (Canada and U.S.)					
females, 18–26 yrs	1255	1640 (73)	—	—	69 (18)
Males, 18–65 yrs	1120	1778 (79)	—	—	86 (18)
(Robinette, Blackwell, Daanen et al., 2002)					
Midwest workers, with shoes and light clothes					
Females	125	1637 (62)	—	—	65 (12)
Males (Marras & Kim, 1993)	384	1778 (73)	—	—	84 (16)
Male miners (Kuenzi & Kennedy, 1993)	105	1803 (65)	—	—	89 (15)
U.S. Army soldiers:					
Females	2208	1629 (64)	852 (35)	515 (26)	62 (8)
Males (Gordon, Churchill, & Clauser, et al., 1989)	1774	1756 (67)	914 (36)	559 (28)	76 (11)
Vietnamese, living in the United States:					
Females	30	1559 (61)	—	—	49
Males (Imrhan, Nguyen, & Nguyen, 1993)	41	1646 (60)	—	—	59

Note: Updated 4 December 2004.

TABLE A.2
Anthropometric Estimated Data in mm of British Adults, 19 to 35 Years of Age

Dimension	Men				Women			
	5th Percentile	50th Percentile	95th Percentile	SD	5th Percentile	50th Percentile	95th Percentile	SD
1. Stature	1625	1740	1855	70	1505	1610	1710	62
2. Eye height, standing	1515	1630	1745	69	1405	1505	1610	61
3. Shoulder height (acromion), standing	1315	1425	1535	66	1215	1310	1405	58
4. Elbow height, standing	1005	1090	1180	52	930	1005	1085	46
5. Hip height (trochanter)	840	920	1000	50	740	810	885	43
6. Knuckle height, standing	690	755	825	41	660	720	780	36
7. Fingertip height, standing	590	655	720	38	560	625	685	38
8. Sitting height	850	910	965	36	795	850	910	35
9. Sitting eye height	735	790	845	35	685	740	795	33
10. Sitting shoulder height (acromion)	540	595	645	32	505	555	610	31
11. Sitting elbow height	195	245	295	31	185	235	280	29
12. Sitting thigh height (clearance)	135	160	185	15	125	155	180	17
13. Sitting knee height	490	545	595	32	455	500	540	27
14. Sitting popliteal height	395	440	490	29	355	400	445	27
15. Shoulder–elbow length	330	365	395	20	300	330	360	17
16. Elbow–fingertip length	440	475	510	21	400	430	460	19
17. Overhead grip reach, sitting	1145	1245	1340	60	1060	1150	1235	53

18. Overhead grip reach, standing	1925	2060	2190	80	1790	1905	2020	71
19. Forward grip reach	720	780	835	34	650	705	755	31
20. Arm length, vertical	720	780	840	36	655	705	760	32
21. Downward grip reach	610	665	715	32	555	600	650	29
22. Chest depth	215	250	285	22	210	250	295	27
23. Abdominal depth, sitting	220	270	325	32	205	255	305	30
24. Buttock–knee depth, sitting	540	595	645	31	520	570	620	30
25. Buttock–popliteal depth, sitting	440	495	550	32	435	480	530	30
26. Shoulder breadth (biacromial)	365	400	430	20	325	355	385	18
27. Shoulder breadth (bideltoid)	420	465	510	28	355	395	435	24
28. Hip breadth, sitting	310	360	405	29	310	370	435	38
29. Span	1655	1790	1925	83	1490	1605	1725	71
30. Elbow span	865	945	1020	47	780	850	920	43
31. Head length	180	195	205	8	165	180	190	7
32. Head breadth	145	155	165	6	135	145	155	6
33. Hand length	175	190	205	10	160	175	190	9
34. Hand breadth	80	85	95	5	70	75	85	4
35. Foot length	240	265	285	14	215	235	255	12
36. Foot breadth	85	95	110	6	80	90	100	6
37. Weight (kg)	55	75	94	12	44	63	81	11

Note: Data estimated in or before 1986. From *Bodyspace: Anthropometry, Ergonomics and the Design of Work* (2nd ed.), by S. Pheasant, 1996. London: Taylor & Francis.

TABLE A.3
Anthropometric Estimated Data in mm of French Adults, 18 to 51 Years of Age

Dimension	Men				Women			
	5th Percentile	50th Percentile	95th Percentile	SD	5th Percentile	50th Percentile	95th Percentile	SD
1. Stature	1638	1747	1855	—	1515	1620	1720	—
2. Eye height, standing	—	—	—	—	—	—	—	—
3. Shoulder height (acromion), standing	1336	1434	1534	—	1240	1331	1433	—
4. Elbow height, standing	—	—	—	—	—	—	—	—
5. Hip height (trochanter)	830	918	989	—	774	844	924	—
6. Knuckle height, standing	—	—	—	—	—	—	—	—
7. Fingertip height, standing	—	—	—	—	—	—	—	—
8. Sitting height	863	918	972	—	814	867	911	—
9. Sitting eye height	764	819	879	—	725	772	822	—
10. Sitting shoulder height (acromion)	—	—	—	—	—	—	—	—
11. Sitting elbow height	—	—	—	—	—	—	—	—
12. Sitting thigh height (clearance)	—	—	—	—	—	—	—	—
13. Sitting knee height	491	533	578	—	446	487	531	—
14. Sitting popliteal height	—	—	—	—	—	—	—	—
15. Shoulder–elbow length	336	365	395	—	305	332	358	—
16. Elbow–fingertip length	438	472	505	—	387	427	459	—
17. Overhead grip reach, sitting	—	—	—	—	—	—	—	—

18. Overhead grip reach (standing)	—	—	—	—	—	—
19. Forward grip reach	—	—	—	—	—	—
20. Arm length, vertical	—	—	—	—	—	—
21. Downward grip reach	—	—	—	—	—	—
22. Chest depth	—	—	—	—	—	—
23. Abdominal depth, sitting	549	595	643	521	569	620
24. Buttock–knee depth, sitting	—	—	—	—	—	—
25. Buttock–popliteal depth, sitting	345	382	415	311	340	372
26. Shoulder breadth (biacromial)	418	457	499	369	410	448
27. Shoulder breadth (bideltoid)	310	342	386	308	346	392
28. Hip breadth, sitting	—	—	—	—	—	—
29. Span	—	—	—	—	—	—
30. Elbow span	—	—	—	—	—	—
31. Head length	146	155	164	140	148	158
32. Head breadth	134	142	151	125	134	143
33. Hand length	176	190	204	158	173	187
34. Hand breadth	79	86	95	69	76	83
35. Foot length	243	264	285	218	237	257
36. Foot breadth	92	101	111	82	91	101
37. Weight (kg)	57	70	88	46	58	72

Note: From personal communication, A. Coblentz, April 22, 1997, regarding ERGODATA taken on 1015 French soldiers (687 males and 328 females).

TABLE A.4
Anthropometric Measured Data in mm of East German Adults, 18 to 59 Years of Age

Dimension	Men				Women			
	5th Percentile	Mean	95th Percentile	SD	5th Percentile	Mean	95th Percentile	SD
1. Stature	1607	1715	1825	66	1514	1608	1707	59
2. Eye height, standing	1498	1601	1705	64	1415	1504	1597	57
3. Shoulder height (acromion), standing	1320	1414	1512	60	1232	1319	1403	53
4. Elbow height, standing	—	—	—	—	—	—	—	—
5. Hip height (trochanter)	—	—	—	—	—	—	—	—
6. Knuckle height, standing	682	748	819	42	643	703	764	37
7. Fingertip height, standing	588	652	717	39	557	616	672	35
8. Sitting height	846	903	958	34	804	854	905	31
9. Sitting eye height	719	775	831	34	684	733	782	30
10. Sitting shoulder height (acromion)	552	601	650	31	517	562	609	29
11. Sitting elbow height	198	244	293	29	190	234	282	28
12. Sitting thigh height (clearance)	126	151	176	15	125	148	175	15
13. Sitting knee height	490	531	575	27	458	497	538	24
14. Sitting popliteal height	410	452	496	26	380	416	455	23
15. Shoulder–elbow length	—	—	—	—	—	—	—	—
16. Elbow–fingertip length	432	465	500	20	394	425	556	19
17. Overhead grip reach, sitting	—	—	—	—	—	—	—	—

Measurement								
18. Overhead grip reach, standing	1975	2121	2267	89	1843	1973	2103	79
19. Forward grip reach	704	763	824	37	650	706	767	35
20. Arm length, vertical	704	762	820	35	650	703	758	33
21. Downward grip reach	—	—	—	—	—	—	—	—
22. Chest depth	—	—	—	—	—	—	—	—
23. Abdominal depth, sitting	—	—	—	—	—	—	—	—
24. Buttock–knee depth, sitting	560	603	648	27	541	585	630	27
25. Buttock–popliteal depth, sitting	444	486	527	25	437	479	521	26
26. Shoulder breadth (biacromial)	365	399	430	20	336	365	393	17
27. Shoulder breadth (bideltoid)	432	471	510	24	393	437	481	27
28. Hip breadth, sitting	334	369	406	22	346	401	460	35
29. Span	1640	1760	1885	75	1503	1616	1735	70
30. Elbow span	833	895	911	39	757	817	881	38
31. Head length	179	190	201	7	170	181	191	6
32. Head breadth	148	158	168	6	141	151	160	6
33. Hand length	174	189	205	9	161	174	189	9
34. Hand breadth	81	88	96	5	71	78	85	4
35. Foot length	243	264	285	13	222	241	260	12
36. Foot breadth	91	102	113	6	83	93	104	6
37. Weight (kg)	—	—	—	—	—	—	—	—

Note: Data measured between 1979 (some 1967) and 1982. From *Anthropologischer Atlas,* by B. Fluegel, H. Greil, and K. Sommer, 1986, Berlin, Germany: Tribuene.

TABLE A.5

Anthropometric Measured Data in mm of Japanese Adults, 18 to 35 Years of Age

Dimension	Men				Women			
	5th Percentile	Mean	95th Percentile	SD	5th Percentile	Mean	95th Percentile	SD
1. Stature	1599	1688	1777	55	1510	1584	1671	50
2. Eye height, standing	1489	1577	1664	53	1382	1460	1541	49
3. Shoulder height (acromion), standing	1291	1370	1454	50	1208	1279	1367	48
4. Elbow height, standing	970	1035	1098	39	909	967	1028	37
5. Hip height (trochanter)	775	834	899	38	730	787	847	35
6. Knuckle height, standing	—	—	—	—	—	—	—	—
7. Fingertip height, standing	600	644	694	30	563	608	652	27
8. Sitting height	859	910	958	30	810	855	902	28
9. Sitting eye height	741	790	837	29	692	733	778	27
10. Sitting shoulder height (acromion)	549	591	633	26	513	551	588	24
11. Sitting elbow height	216	254	292	23	202	236	269	20
12. Sitting thigh height (clearance)	138	156	176	12	130	143	162	10
13. Sitting knee height	475	509	545	22	442	475	508	20
14. Sitting popliteal height	371	402	434	19	345	372	402	17
15. Shoulder–elbow length	307	337	366	18	289	315	339	15
16. Elbow–fingertip length	418	448	479	18	390	416	445	17
17. Overhead grip reach, sitting	—	—	—	—	—	—	—	—

18. Overhead grip reach, standing	—	—	—	—	—	—	—	—
19. Forward grip reach	—	—	—	—	—	—	—	—
20. Arm length, vertical	—	—	—	—	—	—	—	—
21. Downward grip reach	—	—	—	—	—	—	—	—
22. Chest depth	190	217	246	18	190	215	250	19
23. Abdominal depth, sitting	179	208	245	20	161	188	218	17
24. Buttock–knee depth, sitting	530	567	604	23	511	550	586	22
25. Buttock–popliteal depth, sitting	—	—	—	—	—	—	—	—
26. Shoulder breadth (biacromial)	368	395	423	17	346	367	391	14
27. Shoulder breadth (bideltoid)	—	—	—	—	—	—	—	—
28. Hip breadth, sitting	318	349	380	19	331	358	386	17
29. Span	1591	1690	1795	63	1483	1579	1693	62
30. Elbow span	—	—	—	—	—	—	—	—
31. Head length	178	190	203	7	168	177	187	6
32. Head breadth	152	161	171	6	143	151	160	6
33. Hand length	—	—	—	—	—	—	—	—
34. Hand breadth	79	85	91	4	70	75	81	3
35. Foot length	234	251	269	11	217	232	246	9
36. Foot breadth	97	104	111	5	89	96	103	4
37. Weight (kg)	54	66	80	8	45	54	65	6

Note: Data measured in 1988. From *Anthropometry of JASDF Personnel and its Applications for Human Engineering*, by Y. Kagimoto (Ed.), 1990, Tokyo: Aeromedical Laboratory, Air Development and Test Wing JASDF.

TABLE A.6
Anthropometric Measured Data in mm of Russian Factory Workers

Dimension	Men				Women			
	5th Percentile	Mean	95th Percentile	SD	5th Percentile	Mean	95th Percentile	SD
1. Stature [99]								
Ethnic Russians (Moscow)	1636	1736	1837	61	1519	1606	1693	53
Male Prikames, female Uzbeks	1606	1703	1780	59	1491	1580	1668	54
2. Eye height, standing [D19]								
Ethnic Russians (Moscow)	1517	1613	1709	58	1409	1494	1579	52
Male Prikames, female Uzbeks	1504	1589	1697	58	1385	1469	1553	51
3. Shoulder height (acromion), standing [2]								
Ethnic Russians (Moscow)	1329	1425	1521	58	1224	1303	1383	48
Male Prikames, female Uzbeks	1306	1392	1478	52	1204	1284	1365	49
4. Elbow height, standing [D16]								
Ethnic Russians (Moscow)	990	1070	1150	49	910	974	1038	39
Male Prikames, female Uzbeks	979	1039	1110	41	910	974	1038	39
5. Hip height (trochanter) [107]	—	—	—	—	—	—	—	—
6. Knuckle height, standing								
Ethnic Russians (Moscow)	696	763	831	41	—	—	—	—
Male Prikames, female Uzbeks	687	737	800	33	643	695	747	32
7. Fingertip height, standing [D13]								
Ethnic Russians (Moscow)	591	655	718	39	559	606	652	28
Male Prikames, female Uzbeks	586	633	698	32	552	601	650	30
8. Sitting height [93]								
Ethnic Russians (Moscow)	856	909	962	32	799	859	898	30
Male Prikames, female Uzbeks	842	896	949	32	793	845	897	31

Measurement								
9. Sitting eye height [49]								
Ethnic Russians (Moscow)	740	791	843	32	678	729	770	31
Male Prikames, female Uzbeks	728	778	829	31	680	729	777	30
10. Sitting shoulder height (acromion) [3]	—	—	—	—	—	—	—	—
11. Sitting elbow height [48]								
Ethnic Russians (Moscow)	198	241	289	26	188	228	268	24
Male Prikames, female Uzbeks	192	233	274	25	190	229	267	23
12. Sitting thigh height (clearance) [104]								
Ethnic Russians (Moscow)	120	146	173	16	122	145	168	14
Male Prikames, female Uzbeks	119	137	161	13	117	144	182	18
13. Sitting knee height [73]								
Ethnic Russians (Moscow)	508	550	591	25	451	494	536	26
Male Prikames, female Uzbeks	502	541	580	24	441	484	526	26
14. Sitting popliteal height [86]								
Ethnic Russians (Moscow)	413	450	486	22	375	405	435	19
Male Prikames, female Uzbeks	405	440	475	21	358	393	428	21
15. Shoulder–elbow length [91]	—	—	—	—	—	—	—	—
16. Elbow–fingertip length [54]	—	—	—	—	—	—	—	—
17. Overhead grip reach, sitting [D45]								
Ethnic Russians (Moscow)	1181	1268	1353	52	1063	1140	1216	47
Male Prikames, female Uzbeks	1183	1261	1339	48	—	—	—	—
18. Overhead grip reach, standing [D42]	—	—	—	—	—	—	—	—
19. Forward grip reach [D21]								
Ethnic Russians (Moscow)	694	753	811	36	—	—	—	—
Male Prikames, female Uzbeks	672	726	780	33	612	669	726	35
20. Arm length, vertical [D3]	—	—	—	—	—	—	—	—
21. Downward grip reach [D43]	—	—	—	—	—	—	—	—
22. Chest depth [36]								
Ethnic Russians (Moscow)	219	256	310	28	209	252	302	29

TABLE A.6 (continued)
Anthropometric Measured Data in mm of Russian Factory Workers

Dimension	Men				Women			
	5th Percentile	Mean	95th Percentile	SD	5th Percentile	Mean	95th Percentile	SD
Uzbeks (Tashkent)	214	246	284	21	200	234	298	27
23. Abdominal depth, sitting [1]	—	—	—	—	—	—	—	—
24. Buttock–knee depth, sitting [26]								
Ethnic Russians (Moscow)	550	601	652	31	519	565	610	28
Male Prikames, female Uzbeks	539	583	627	27	506	559	613	33
25. Buttock–popliteal depth, sitting [27]								
Ethnic Russians (Moscow)	455	504	554	30	432	479	527	29
Male Prikames, female Uzbeks	448	488	528	24	426	477	529	31
26. Shoulder breadth (biacromial) [10]								
Ethnic Russians (Moscow)	381	408	426	17	332	361	390	18
Male Prikames, female Uzbeks	362	392	422	18	316	345	374	18
27. Shoulder breadth (bideltoid) [12]								
Ethnic Russians (Moscow)	414	450	486	22	378	418	458	24
Male Prikames, female Uzbeks	412	442	480	20	378	388	441	27
28. Hip breadth, sitting [66]								
Ethnic Russians (Moscow)	313	348	394	24	333	379	425	23
Male Prikames, female Uzbeks	313	344	385	20	321	365	423	30
29. Span [98]								
Ethnic Russians (Moscow)	1665	1787	1909	74	1492	1624	1756	80
Male Prikames, female Uzbeks	1645	1758	1871	69	1473	1587	1701	69

30. Elbow span.								
Ethnic Russians (Moscow)	875	936	998	37	810	868	927	36
Male Prikames, female Uzbeks	843	913	982	42	751	819	888	41
31. Head length [62]	—	—	—	—	—	—	—	—
32. Head breadth [60]	—	—	—	—	—	—	—	—
33. Hand length [59]								
Ethnic Russians (Moscow), 18–50 years	169	184	195	8	153	165	178	8
Female Uzbeks, 18–45 years	—	—	—	—	153	166	179	8
34. Hand breadth [57]								
Ethnic Russians (Moscow), 18–50 years	81	88	95	4	73	79	86	4
Female Uzbeks, 18–50years	—	—	—	—	71	78	86	5
35. Foot length [51]								
Ethnic Russians (Moscow)	251	270	289	12	221	241	262	12
Male Prikames, female Uzbeks	241	260	279	12	217	235	254	11
36. Foot breadth [50]								
Ethnic Russians (Moscow)	90	98	105	5	83	91	99	5
Male Prikames, female Uzbeks	92	100	109	5	74	88	103	9
37. Weight (kg)								
Ethnic Russians (Moscow)	59	72	92	10	49	61	76	8
Ethnic Russians (Moscow), 30–50 years	59	76	97	11	50	66	87	11
Male Prikames, female Uzbeks	57	68	86	8	45	58	76	9
Male Prikames, female Uzbeks, 30–45 years	58	73	88	9	49	68	92	13

Note: The samples of ethnic Russians are 192 males between 18 and 29 years old and 205 females between 20 and 29 years old, all from Moscow. Prikames comprise a population group of combined ethnic Russians, Tartars, Chuvashis, and others, numbering altogether 240 male persons between 20 and 29 years of age. Their comparison group consists of 301 female Uzbeks between 18 and 29 years of age. Exceptions are stated. Measurements were taken between 1985 and 1986. The measurements are similar to those defined by Gordon, Churchill, Clauser, et al. (1989) with their reference numbers in brackets. From *Anthropo-ergonomic Atlas* (in Russian, ISBN 5-211-04102-X), by A. N. Strokina and B. A. Pakhomova, 1999. Moscow, Russia: Moscow State University Publishing House.

TABLE A.7

Anthropometric Measured Data in mm of Russian Coal Miners and Ethnic Asian Agricultural Workers

Dimension	Men				Women			
	5th Percentile	Mean	95th Percentile	SD	5th Percentile	Mean	95th Percentile	SD
1. Stature [99]								
Male coal miners	1701	1801	1900	61	—	—	—	—
Male mechanics, female pig herders	1608	1704	1794	58	1497	1588	1678	55
2. Eye height, standing [D19]								
Male coal miners	1534	1628	1721	57	—	—	—	—
Male mechanics, female pig herders	1490	1585	1679	57	1395	1480	1565	62
3. Shoulder height (acromion), standing [2]								
Male coal miners	—	—	—	—	—	—	—	—
Male mechanics, female pig herders	1301	1391	1481	55	1247	1330	1418	53
4. Elbow height, standing [D16]								
Male coal miners	1017	1094	1191	47	—	—	—	—
Male mechanics, female pig herders	974	1041	1108	41	915	982	1048	40
5. Hip height (trochanter) [107]								
Male coal miners	—	—	—	—	—	—	—	—
6. Knuckle height, standing								
Male coal miners	696	763	831	41	—	—	—	—
Male mechanics, female pig herders	683	742	801	36	645	704	763	36
7. Fingertip height, standing [D13]								
Male coal miners	636	704	771	41	—	—	—	—
Male mechanics, female pig herders	574	629	685	34	550	609	669	36
8. Sitting height [93]								
Male coal miners	923	978	1032	33	—	—	—	—
Male mechanics	851	902	954	31	—	—	—	—

Measurement								
9. Sitting eye height [49]								
Male coal miners	754	806	856	31	—	—	—	—
Male mechanics	738	788	838	30	—	—	—	—
10. Sitting shoulder height (acromion) [3]	—	—	—	—	—	—	—	—
11. Sitting elbow height [48]								
Male coal miners	200	248	295	29	—	—	—	—
Male mechanics	183	229	274	28	—	—	—	—
12. Sitting thigh height (clearance) [104]								
Male coal miners	147	171	197	15	—	—	—	—
Male mechanics	116	139	162	14	—	—	—	—
13. Sitting knee height [73]								
Male coal miners	531	572	634	25	—	—	—	—
Male mechanics	486	530	568	25	—	—	—	—
14. Sitting popliteal height [86]								
Male coal miners	411	462	514	31	—	—	—	—
Male mechanics	396	429	462	20	—	—	—	—
15. Shoulder–elbow length [91]	—	—	—	—	—	—	—	—
16. Elbow–fingertip length [54]	—	—	—	—	—	—	—	—
17. Overhead grip reach, sitting [D45]								
Male coal miners	1217	1315	1412	60	—	—	—	—
Male mechanics	1186	1256	1326	43	—	—	—	—
18. Overhead grip reach, standing [D42]	—	—	—	—	—	—	—	—
19. Forward grip reach [D21]								
Male coal miners	684	737	790	32	—	—	—	—
Male mechanics	—	—	—	—	—	—	—	—
20. Arm length, vertical [D3]	—	—	—	—	—	—	—	—
21. Downward grip reach [D43]	—	—	—	—	—	—	—	—
22. Chest depth [36]								
Male coal miners	302	345	389	27	—	—	—	—
Male mechanics, female pig herders	211	245	280	21	198	267	336	42

TABLE A.7 (continued)
Anthropometric Measured Data in mm of Russian Coal Miners and Ethnic Asian Workers

Dimension	Men				Women			
	5th Percentile	Mean	95th Percentile	SD	5th Percentile	Mean	95th Percentile	SD
23. Abdominal depth, sitting [1]	—	—	—	—	—	—	—	—
24. Buttock–knee depth, sitting [26]								
Male coal miners	—	—	—	—	—	—	—	—
Male mechanics	538	594	651	34	—	—	—	—
25. Buttock–popliteal depth, sitting [27]								
Male coal miners	411	?	?	?	—	—	—	—
Male mechanics	444	499	554	34	—	—	—	—
26. Shoulder breadth (biacromial) [10]								
Male coal miners	—	—	—	—	—	—	—	—
Male mechanics, female pig herders	359	389	418	18	321	351	381	18
27. Shoulder breadth (bideltoid) [12]								
Male coal miners	495	542	589	29	—	—	—	—
Male mechanics, female pig herders	393	432	472	24	351	412	473	37
28. Hip breadth, sitting [66]								
Male coal miners	313	348	394	24	—	—	—	—
Male mechanics	315	350	385	21	—	—	—	—
29. Span [98]								
Male coal miners	1665	1787	1909	74	—	—	—	—
Male mechanics, female pig herders	1646	1746	1846	61	1498	1610	1722	68

30. Elbow span								
Male coal miners	875	936	998	37	—	—	—	—
Male mechanics, female pig herders	835	905	976	43	767	840	912	44
31. Head length [62]	—	—						
32. Head breadth [60]	—	—						
33. Hand length [59]								
Male coal miners	169	184	195	8	—	—	—	—
Male mechanics*	172	188	204	10	—	—	—	—
34. Hand breadth [57]								
Male coal miners	81	88	95	4	—	—	—	—
Male mechanics*, female pig herders	83	91	98	5	74	80	86	4
35. Foot length [51]								
Male coal miners	251	270	289	12	—	—	—	—
Male mechanics	239	259	278	12	—	—	—	—
36. Foot breadth [50]								
Male coal miners	90	98	105	5	—	—	—	—
Male mechanics	91	98	111	8	—	—	—	—
37. Weight (kg)								
Male coal miners	51	64	77	8	—	—	—	—
Male mechanics								

Note: The sample of ethnic Russian coal miners consists of 150 males between 20 and 50 years old from the Ukraine and Belarussia. The agricultural mechanics comprise ethnic Uzbeks, Tajiks, and Turkmens, together 520 male persons between 20 and 29 (*20 and 50) years of age. Their comparison group consists of 246 female pig herders, also between 20 and 29 years of age, from the Saratov, Riazan, Kalinin, and Novosibirsk areas. Measurements were taken in the 1980s. The measurements are similar to those defined by Gordon, Churchill, Clauser, et al. (1989) with their reference numbers in brackets. From *Anthropo-ergonomic Atlas* (in Russian, ISBN 5-211-04102-X), by A. N. Strokina and B. A. Pakhomova, 1999, Moscow, Russia: Moscow State University Publishing House.

TABLE A.8
Anthropometric Measured Data in mm of Russian Students, 18 to 22 Years of Age

Dimension	Men				Women			
	5th Percentile	Mean	95th Percentile	SD	5th Percentile	Mean	95th Percentile	SD
1. Stature [99]								
Ethnic Russians (Moscow)	1664	1757	1849	56	1542	1637	1731	57
Uzbeks (Tashkent)	1615	1700	1786	52	1498	1578	1658	52
2. Eye height, standing [D19]								
Ethnic Russians (Moscow)	1547	1637	1728	55	1433	1526	1618	57
Uzbeks (Tashkent)	1496	1581	1665	51	1387	1463	1538	46
3. Shoulder height (acromion), standing [2]								
Ethnic Russians (Moscow)	1351	1440	1529	54	1245	1334	1422	54
Uzbeks (Tashkent)	1313	1391	1469	48	1217	1284	1371	47
4. Elbow height, standing [D16]								
Ethnic Russians (Moscow)	1004	1083	1162	48	941	1010	1080	42
Uzbeks (Tashkent)	985	1042	1099	35	909	970	1031	37
5. Hip height (trochanter) [107]	—	—	—	—	—	—	—	—
6. Knuckle height, standing								
Ethnic Russians (Moscow)	710	773	836	39	676	731	786	34
Uzbeks (Tashkent)	676	734	792	35	632	687	742	33
7. Fingertip height, standing [D13]								
Ethnic Russians (Moscow)	508	668	729	37	582	635	687	32
Uzbeks (Tashkent)	579	635	691	34	546	599	652	32
8. Sitting height [93]								
Ethnic Russians (Moscow)	860	912	964	32	806	859	911	32
Uzbeks (Tashkent)	858	905	952	29	793	839	885	28

9. Sitting eye height [49]								
Ethnic Russians (Moscow)	737	790	844	33	694	742	790	29
Uzbeks (Tashkent)	737	784	830	28	676	723	771	29
10. Sitting shoulder height (acromion) [3]	—	—	—	—	—	—	—	—
11. Sitting elbow height [48]								
Ethnic Russians (Moscow)	202	243	284	25	196	236	275	24
Uzbeks (Tashkent)	186	229	272	26	191	229	267	23
12. Sitting thigh height (clearance) [104]								
Ethnic Russians (Moscow)	122	151	179	18	126	148	172	14
Uzbeks (Tashkent)	120	143	165	14	114	142	170	17
13. Sitting knee height [73]								
Ethnic Russians (Moscow)	520	562	603	25	487	527	567	24
Uzbeks (Tashkent)	494	531	569	23	446	487	528	25
14. Sitting popliteal height [86]								
Ethnic Russians (Moscow)	429	468	508	24	386	423	461	23
Uzbeks (Tashkent)	400	430	460	18	366	398	430	20
15. Shoulder–elbow length [91]	—	—	—	—	—	—	—	—
16. Elbow–fingertip length [54]	—	—	—	—	—	—	—	—
17. Overhead grip reach, sitting [D45]								
Ethnic Russians (Moscow)	1199	1276	1354	47	1094	1169	1244	46
Uzbeks (Tashkent)	1193	1256	1319	38	1085	1152	1219	41
18. Overhead grip reach, standing [D42]	—	—	—	—	—	—	—	—
19. Forward grip reach [D21]								
Ethnic Russians (Moscow)	697	759	821	38	641	702	763	37
Uzbeks (Tashkent)	686	745	803	36	609	673	737	39
20. Arm length, vertical [D3]	—	—	—	—	—	—	—	—
21. Downward grip reach [D43]	—	—	—	—	—	—	—	—
22. Chest depth [36]								
Ethnic Russians (Moscow)	207	245	312	20	209	242	256	21
Uzbeks (Tashkent)	211	244	276	20	200	233	265	20

TABLE A.8 (continued)
Anthropometric Measured Data in mm of Russian Students, 18 to 22 Years of Age

Dimension	Men				Women			
	5th Percentile	Mean	95th Percentile	SD	5th Percentile	Mean	95th Percentile	SD
23. Abdominal depth, sitting [1]	—	—	—	—	—	—	—	—
24. Buttock–knee depth, sitting [26]								
Ethnic Russians (Moscow)	561	610	660	30	536	584	631	29
Uzbeks (Tashkent)	541	595	648	33	515	564	612	30
25. Buttock–popliteal depth, sitting [27]								
Ethnic Russians (Moscow)	476	517	557	25	446	496	540	29
Uzbeks (Tashkent)	459	504	550	28	423	472	520	29
26. Shoulder breadth (biacromial) [10]								
Ethnic Russians (Moscow)	369	397	425	17	334	360	386	16
Uzbeks (Tashkent)	349	377	404	17	320	347	373	16
27. Shoulder breadth (bideltoid) [12]								
Ethnic Russians (Moscow)	416	458	492	23	377	412	446	21
Uzbeks (Tashkent)	409	438	466	17	352	381	410	17
28. Hip breadth, sitting [66]								
Ethnic Russians (Moscow)	323	362	410	23	334	372	411	23
Uzbeks (Tashkent)	316	349	381	20	329	364	399	21
29. Span [98]								
Ethnic Russians (Moscow)	1671	1782	1893	68	1516	1640	1763	75
Uzbeks (Tashkent)	1640	1747	1855	66	1461	1579	1698	72

30. Elbow span								
Ethnic Russians (Moscow)	874	935	995	37	808	870	933	38
Uzbeks (Tashkent)	842	909	976	41	781	837	894	34
31. Head length [62]	—	—	—	—	—	—	—	—
32. Head breadth [60]	—	—	—	—	—	—	—	—
33. Hand length [59]								
Ethnic Russians (Moscow)	174	188	202	9	155	168	182	8
Uzbeks (Tashkent)	175	188	201	8	—	—	—	—
34. Hand breadth [57]								
Ethnic Russians (Moscow)	80	87	95	5	71	76	82	3
Uzbeks (Tashkent)	82	89	96	4	73	79	87	4
35. Foot length [51]								
Ethnic Russians (Moscow)	247	266	286	12	222	239	256	11
Uzbeks (Tashkent)	242	260	279	11	220	237	254	10
36. Foot breadth [50]								
Ethnic Russians (Moscow)	87	97	107	6	92	88	95	4
Uzbeks (Tashkent)	85	96	107	7	81	90	98	5
37. Weight (kg)								
Ethnic Russians (Moscow)	57	71	85	9	49	60	73	7
Uzbeks (Tashkent)	53	65	76	7	45	56	68	7

Note: The measured sample consisted of 166 male and 207 female Russians from Moscow and 150 male and 164 female Uzbeks from Tashkent. Measurements were taken between 1984 and 1986. The measurements are similar to those defined by Gordon, Churchill, Clauser, et al. (1989) with their reference numbers in brackets. From *Anthropo-ergonomic Atlas* (in Russian, ISBN 5-211-04102-X), by A. N. Strokina and B. A. Pakhomova, 1999, Moscow, Russia: Moscow State University Publishing House.

TABLE A.9
Anthropometric Measured Data in mm of Chinese (Taiwan) Adults, 25 to 34 Years of Age

Dimension	Men				Women			
	5th Percentile	Mean	95th Percentile	SD	5th Percentile	Mean	95th Percentile	SD
1. Stature [99]	1608	1705	1801	59	1485	1572	1659	53
2. Eye height, standing [D19]	—	—	—	—	—	—	—	—
3. Shoulder height (acromion), standing [2]	1309	1396	1484	53	1204	1285	1367	50
4. Elbow height, standing [D16]	993	1059	1126	40	915	978	1040	38
5. Hip height (trochanter) [107]	780	860	939	48	735	802	869	41
6. Knuckle height, standing	705	757	809	32	653	708	762	33
7. Fingertip height, standing [D13]	610	659	708	30	566	618	670	32
8. Sitting height [93]	861	910	959	30	794	846	898	32
9. Sitting eye height [49]	742	791	839	29	681	732	783	31
10. Sitting shoulder height (acromion) [3]	560	602	645	26	516	561	605	27
11. Sitting elbow height [48]	226	264	303	24	211	252	294	25
12. Sitting thigh height (clearance) [104]	—	—	—	—	—	—	—	—
13. Sitting knee height [73]	474	521	569	29	431	471	510	24
14. Sitting popliteal height [86]	380	411	442	19	350	379	408	18
15. Shoulder–elbow length [91]	308	338	369	19	280	309	339	18
16. Elbow–fingertip length [54]	382	427	472	27	339	384	429	27
17. Overhead grip reach, sitting [D45]	1128	1208	1289	49	1033	1105	1177	44

18. Overhead grip reach, standing [D42]	1872	2002	2133	79	1721	1831	1942	67
19. Forward grip reach [D21]	650	710	770	36	597	651	705	33
20. Arm length, vertical [D3]	684	738	793	33	618	669	720	31
21. Downward grip reach [D43]	—	—	—	—	—	—	—	—
22. Chest depth [36]	187	217	248	19	182	213	244	19
23. Abdominal depth, sitting [1]	—	—	—	—	—	—	—	—
24. Buttock–knee depth, sitting [26]	507	558	608	31	487	530	572	26
25. Buttock–popliteal depth, sitting [27]	—	—	—	—	—	—	—	—
26. Shoulder breadth (biacromial) [10]	323	369	415	28	282	324	366	25
27. Shoulder breadth (bideltoid) [12]	422	460	499	23	367	406	445	24
28. Hip breadth, sitting [66]	315	360	404	27	316	353	390	23
29. Span [98]	1625	1738	1852	69	1469	1571	1672	62
30. Elbow span	820	894	968	45	737	801	866	39
31. Head length [62]	185	197	209	7	176	187	198	6
32. Head breadth [60]	154	167	181	8	146	161	175	9
33. Hand length [59]	168	183	199	10	154	167	181	8
34. Hand breadth [57]	77	86	94	5	68	75	82	4
35. Foot length [51]	—	—	—	—	—	—	—	—
36. Foot breadth [50]	—	—	—	—	—	—	—	—
37. Weight (kg)	53	67	81	9	40	52	64	7

Note: Measurements were taken 1996–2000 on nearly 1200 civilians. The measurments are similar to those defined by Gordon, Churchill, Clauser et al. (1989) with their reference numbers in brackets. From *Anthropometric Data Book of the Chinese People in Taiwan*, by M. J. J. Wang, E. M. Y. Wang, and Y. C. Lin, 2002. Hsinchu, ROC: The Ergonomics Society of Taiwan.

TABLE A.10
Anthropometric Measured Data in mm of U.S. Army Personnel, 17 to 51 Years of Age

Dimension	Men				Women			
	5th Percentile	Mean	95th Percentile	SD	5th Percentile	Mean	95th Percentile	SD
1. Stature [99]	1647	1756	1867	67	1528	1629	1737	64
2. Eye height, standing [D19]	1528	1634	1743	66	1415	1516	1621	63
3. Shoulder height (acromion), standing [2]	1342	1443	1546	62	1241	1334	1432	58
4. Elbow height, standing [D16]	995	1073	1153	48	926	998	1074	45
5. Hip height (trochanter) [107]	853	928	1009	48	789	862	938	45
6. Knuckle height, standing	—	—	—	—	—	—	—	—
7. Fingertip height, standing [D13]	591	653	716	40	551	610	670	36
8. Sitting height [93]	855	914	972	36	795	852	910	35
9. Sitting eye height [49]	735	792	848	34	685	739	794	33
10. Sitting shoulder height (acromion) [3]	549	598	646	30	509	556	604	29
11. Sitting elbow height [48]	184	231	274	27	176	221	264	27
12. Sitting thigh height (clearance) [104]	149	168	190	13	140	160	180	12
13. Sitting knee height [73]	514	559	606	28	474	515	560	26
14. Sitting popliteal height [86]	395	434	476	25	351	389	429	24
15. Shoulder–elbow length [91]	340	369	399	18	308	336	365	17
16. Elbow–fingertip length [54]	448	484	524	23	406	443	483	23
17. Overhead grip reach, sitting [D45]	1221	1310	1401	55	1127	1212	1296	51
18. Overhead grip reach, standing [D42]	1958	2107	2260	92	1808	1947	2094	87
19. Forward grip reach [D21]	693	751	813	37	632	686	744	34

20. Arm length, vertical [D3]	729	790	856	39	662	724	788	38
21. Downward grip reach [D43]	612	666	722	33	557	700	664	33
22. Chest depth [36]	210	243	280	22	209	239	279	21
23. Abdominal depth, sitting [1]	199	236	291	28	185	219	271	26
24. Buttock–knee depth, sitting [26]	569	616	667	30	542	589	640	30
25. Buttock–popliteal depth, sitting [27]	458	500	546	27	440	482	528	27
26. Shoulder breadth (biacromial) [10]	367	397	426	18	333	363	391	17
27. Shoulder breadth (bideltoid) [12]	450	492	535	26	397	433	472	23
28. Hip breadth, sitting [66]	329	367	412	25	343	385	432	27
29. Span [98]	1693	1823	1960	82	1542	1672	1809	81
30. Elbow span	—	—	—	—	—	—	—	—
31. Head length [62]	185	197	209	7	176	187	198	6
32. Head breadth [60]	143	152	161	5	137	144	153	5
33. Hand length [59]	179	194	211	10	165	181	197	10
34. Hand breadth [57]	84	90	98	4	73	79	86	4
35. Foot length [51]	249	270	292	13	224	244	265	12
36. Foot breadth [50]	92	101	110	5	82	90	98	5
37. Weight (kg)	62	79	98	11	50	62	77	8

Note: Measurements were taken in 1987–1988 on U.S. Army soldiers, 1774 men and 2208 women. Kroemer, Kroemer, and Kroemer (1997b) reasoned that these data were good estimates for U.S. civilians as well, but given the trend of increasing obesity, the current civilians' weights (and related dimensions, especially abdominal depths) are probably considerably higher than listed on line 37. Data, including numbers in brackets, were taken from *1988 Anthropometric Survey of U.S. Army Personnel. Summary Statistics Interim Report* (Technical Report NATICK/TR-89/027), by C. C. Gordon, T. Churchill, C. E. Clauser, B. Bradtmiller, J. T. McConville, I. Tebbetts, and R. A. Walker, 1989, Natick, MA: United States Army Natick Research, Development and Engineering Center.

Index

A

Activities of daily living (ADL), 47, 48, 49, 50, 55, 147, 148
Acuity, 20
ADL, 47, 48, 49, 50, 55, 147, 148
Adolescents, 175, 176, 180, 182, 183, 189, 190
Aging, 128
Aging-related changes, 130–140
Anthropometric data, 8, 60, 62, 69
   calculating data, 98
Anthropometric techniques, 33, 34, 52, 86, 87, 99
Anthropometry, 7, 52, 130, 176
   3-D, 33, 34, 52, 86, 87, 99
   aging persons, 130–131
   arm and hand, 89–91, 95, 97
   children, 66, 175–193
   correlations, 61
   disabled persons, 34–35, 117
   head, 88, 90, 96
   leg and foot, 93, 97
   measurements, 92, 166, 181
   measuring techniques, 33, 34, 52, 86, 87, 99, 166
   pregnant women, 33, 79, 86, 165
   reach, 94–96
   sitting height, 90
   standing height, stature, 88
   trunk, 88–89, 91, 95–96
   surveys, 8, 48, 124, 179
   weight, 97, 103
Anthropometry of adults by country/region
   Algeria, 222
   Australia, 222
   Brazil, 222
   Britain, 131, 165, 189, 191, 226
   China, 222, 246
   Egypt, 222
   France, 228
   Germany, 222, 230
   India, 222, 223
   Indonesia, 223
   Iran, 223
   Ireland, 223
   Italy, 223
   Jamaica, 223
   Japan, 3, 7, 165, 232

   Korea, 223
   Malaysia, 223
   Netherlands, 34, 66, 131, 223
   Saudi Arabia, 223
   Singapore, 223
   Sri Lanka, 223
   Sudan, 223
   Russia, 178, 179, 223, 234–245
   Taiwan, 181
   Thailand, 223
   Turkey, 223
   United Kingdom, 131, 165, 191, 226
   USA, 1, 7, 34, 61–63, 130, 165, 225, 248
Anthropometry of children by country/region
   Algeria, 179
   Belgium, 179
   China, 179–183
   Germany, 165, 167, 179–183
   Holland, 34, 66, 131, 179–183
   Hungary, 179
   France, 179–183
   India, 179
   Japan, 3, 7, 165, 179–183
   Korea, 179
   Netherlands, 34, 66, 131, 179–183
   New Zealand, 179
   Norway, 179
   Nubia, 179
   Portugal, 179–183
   Saudi Arabia, 179
   Sweden, 179–183, 191
   Turkey, 179
   Russia, 178, 179
   United Kingdom, 131, 165, 179–183, 191
   USA, 175–193
Assessing auditory capabilities, 41
Assessing complex capabilities, 46
   biomechanics, 52
   gerontology, 50
   industrial engineering, 53
   intrinsic performance elements (IPEs), 48
   medicine, 27, 42, 50
   physiology, 27, 42, 51
   possible solutions, 55
   psychology, 51
   rehabilitation engineering, 53, 117
   sports sciences, 52